Quality assurance in construction

Proceedings of the conference *Quality assurance for the chief executive,* organized by the Institution of Civil Engineers and held in London on 14 February 1989

Thomas Telford, London

Co-sponsored by the Chartered Institute of Building

Organizing Committee: J. L. Ashford (Chairman),
L. J. L. Carvalho, R. H. Courtier, Dr W. G. Curtin, M. J. Lloyd,
Professor R. McCaffer, D. Culverwell

British Library Cataloguing in Publication Data
Quality assurance in construction
   1.Construction industries. Quality assurance
   I. Institution of Civil Engineers
   624'.068'5

   ISBN 0-7277-1536-4

Published for the Institution of Civil Engineers by Thomas Telford Ltd, Telford
House, 1 Heron Quay, London E14 9XF.

Printed and bound in Great Britain by Billing and Sons Ltd, Worcester.

Contents

Keynote address

Sir Clifford CHETWOOD, George Wimpey plc

Quality problems are not a new phenomenon. They have always existed. In the past, the cause of much defective work was ignorance. There is no longer that excuse because enough technical knowledge exists to enable high standards of work to be maintained. The difficulty is to make sure that this know-how is effectively applied where and when it is needed. This is a management function, and management starts at the top. So what is the role of the chief executive in managing quality? Three key tasks exist which only he can perform.

The first of these is the establishment of a corporate ethos. What is the firm's purpose? What are its beliefs? How does the company expect its people to behave? Questions such as these invite a cynical response. Surely a company exists to make money, does it not? Obviously, companies which do not make a profit will go out of business. But profit by itself is not enough. For a start, there is the question of time-scale. Is prosperity to be short-term or long-term? Does the organization want to maintain an on-going business on a basis of integrity and compliance with the rules, or does it wish to pursue short-term profits by tolerating bad practices and short-changing its clients? The ethics of a poorly managed contractor faced with the overriding imperative of paying the wages at the end of the week can be expected to differ significantly from those of a mature business, dependent on the continued loyalty of its work-force and with a reputation to maintain. These are matters of considerable interest to potential clients who need to know the kind of organization they will be dealing with. They are also of interest to the employees. If profit is management's only objective, it can not complain if the work-force follows its example and works only to maximize its own income. This can be a short cut to disaster.

To achieve long-term prosperity a company has to satisfy three parties. It has to generate profits to satisfy the shareholders and provide funds for future investment. It has to satify the needs of clients for works completed to specification, within budget and on time. Last but not least, it has to satisfy the needs of its employees, including their training, safety and material rewards. The corporate ethos must concentrate and direct the company's resources towards the satisfaction of these three needs. They are like the three legs of a tripod. Remove one leg and the whole will fall.

The second key task of the chief executive is to set the system in motion. He has to select the right people and establish an organization structure in which they can function effectively. He has to make sure that tasks are defined and responsibilities delegated. He has to prevail on line managers to devise and document the management systems which are to be followed. He has to instil attitudes of pride and self-respect whereby people can become accountable for the quality of their work without constant supervision. Many of these concepts are in conflict with the traditions of the construction industry. The task is therefore difficult. Long cherished customs have to be abandoned. Skeletons have to be brought out of cupboards and vested interests have to be challenged. Inevitably many people will find the process disturbing and uncomfortable. When the going gets rough the chief executive must maintain his purpose and thus the momentum.

The third key task of the chief executive is the provision of leadership. Modern techniques of quality management require a change in attitudes. Old habits die hard, and one which lingers longer than most is the belief that each party in the construction team can concentrate on minimizing its own costs while relying on others to protect and satisfy the client. Such attitudes are incompatible with the new realities and they have to change if progress is to be made.

The implication of this is that people at all levels are being asked to abandon the behaviour styles that got them where they are, and to adopt a style which, in many respects, is diametrically opposed to the models of successful behaviour with which they have grown up. The only force which will cause whole organizations to make this leap into the future is coherent and compelling leadership.

The kind of leadership needed requires people who know with absolute certainty what they believe in, what their objectives are and how they are going to achieve these objectives. A chief executive who is not prepared to take the lead, openly and conspicuously, will not succeed in his task. It follows, therefore, that the first person the chief executive must convince to make the quality journey is himself. Only then can he in turn convince the cynics and sceptics that the quality management system is for real, and not just another layer of wallpaper to cover up the cracks.

Is it worth it? I am convinced that it is, for just three reasons. Management is inseparable from control. Only when control is established and demonstrable can the chief executive face the future with any confidence. To demonstrate financial control a company has financial audits. Quality audits are complementary to financial audits, and if one is to have a complete health check of a business it is necessary to have both financial assurance and quality assurance. If profits have been achieved at the expense of quality, they are likely to prove illusory. Companies seeking long-term prosperity need reliable information on the quality of their product as well as on their financial status.

The second important virtue of a quality system is that it provides a mechanism for managing change. The one aspect of the construction industry which can be predicted with certainty is its continuing unpredictability. It has to be accepted that a state of change is not a passing phase, it is normality. When things stop changing, it means either bankruptcy or death. Management systems must therefore act as the instruments of change and not a means of preserving the status quo.

The enterprises which will survive and prosper into the twenty-first century will be those who have learnt to perform a difficult double-act. First they must organize themselves so that their products constantly, consistently and without deviation, meet the requirements of the customers. At the same time, they will have continually to transform their management processes, systems and structures to respond effectively to the ever-changing commercial environment. Companies will have to develop a capability for stable self-transformation. They will have to learn to manage their own change without, at the same time, throwing themselves into turmoil.

The more open style of management and the feedback of information generated by the audit process enables senior management to see the organization as it really is and not as their intermediate subordinates might like them to see it. They are then able to carry out a continual fine tuning of the management system as it becomes necessary and not wait until it deteriorates to a point at which radical surgery is the only cure. It is exceedingly difficult, if not impossible, to change a management system which is not capable of identificaton and under control.

The third benefit of a quality system ties in with the starting point, the corporate ethos. The best people, those whom the industry wants to attract and retain, like to work for a company they can respect. The quality of young people now entering the industry is impressive. They have a strong streak of idealism. They want to use the training they have received to do a useful job of work of which they can be proud. If they are not provided with an opportunity to feel this pride, they will walk away from the industry, leaving only those with less admirable ambitions. The industry cannot afford to let this happen. The number of young people entering industry is about to fall drastically. To attract a sufficient quantity of the right calibre, the negative image which the construction industry all too often reflects must be shed, the fact that it exists to serve society, not to exploit it, must be demonstrated.

3

An historical perspective

R. McCAFFER, BSc, PhD, MICE, MCIOB, Loughborough University of Technology

The construction industry's interests in obtaining quality and in purchasing goods or products fit for their purpose is at an all-time high. The occurrences of even modest failures are well recorded, and in construction the contractors and designers are publicly tried in the semi-technical press. The large failures, of course, lead to enquiries where the evidence is more carefully assembled and considered and the judgement more balanced. In the wider community this is marked by the emergence of more rigorous consumer and trade practices legislation. This interest has either led to or is derived from an expectation that the suppliers can get it right first time, every time. In manufacturing we have seen other countries achieve quality standards deemed impossible in the UK, only later to be forced to emulate them. It has, in these industries, required a change of philosophy. The philosopy of inspection of completed products (or parts) held for so long in the UK as a means of providing quality has proved inadequate. The ethos of quality assurance is clear: quality cannot be inspected in, it must be planned for and built in. The responsibility of each and every producer in the manufacturing system even down to individual level must be clear and accountable. Quality assurance is, as much as anything, an issue of attitude. When the responsibility for checking quality is with the inspector the responsibility is perceived to be removed from the producer. This leads to failures by the producers and uncheckable burdens being placed on inspectors. Yet much of construction's practices are still based on the philosophy of inspection.

Evidence of the construction industry's inspection mentality not working can be obtained from even modest efforts of scrutinizing the technical press where examples abound and a catalogue of failures, large and small, occurring over the years can easily be compiled

    (a)   the high rise blocks of the 1960s of which Ronan Point is the most famous

    (b)   the many failures derived from the use of high alumina cement, such as the Camden School for Girls in 1973

(c)   cracks in the M1 bridges in Derbyshire
(d)   the Tay Road Bridge found to have extensive cracking
       in 1984.

A list of almost any length could be compiled.

But is this new or has the industry always been that bad?
The problem with placing today's quality performance in an
historical setting is that history, by and large, records
only the major catastrophes. The many minor ones went
unrecorded.

I was particularly keen to cite the cracking in the Tay
Road Bridge, 18 years after its opening, as an example of a
modern failure. The bridge still stands and offers a
contrast between a failure of today and a failure in 1879
when the 13 central spans of the first Tay Rail Bridge
collapsed with engine 224, its train and its passengers
after only 19 months of operation. The events surrounding
this tragedy are well recorded and stand as a salutory
lesson in the failure of quality management, for example

(a)   inappropriate materials:the iron used to cast the
       columns was of poor quality
(b)   lack of skill:the moulders had not used that type of
       iron before and found it difficult to work with
(c)   poor work covered up:scabbing on the columns, where
       mould sand had mixed with the iron, was chiselled off
       and filled, but not with iron
(d)   columns not to specification:lugs were missing and
       had to be burnt on, if there was iron on which to
       attach them; bolt holes were cast conically not
       cylindrically, leading to loose fitting and
       undersized bolts being used
(e)   inadequate construction:bracing bars were not
       tightened up, and lessons were not learned from a
       collapse during construction
(f)   inadequate maintenance:the bridge superintendent did
       not have experience of cast iron work, being skilled
       in masonry; the significance of cracks went unnoticed
       and were painted over and loose bracing bars were
       ignored.

The reports presented in court in 1880 were unequivocal,
quote: 'the bridge was badly designed, badly constructed and
badly maintained.' Failure can be seen to fit into the
categories now so familiar in quality assurance literature:
lack of skill, lack of care, lack of knowledge, lack of
accountability, coupled with poor management, and with both
workforce and designer not understanding the materials. Is
the industry now just as bad? I think not and I believe the
drive to improve quality is one of refinement.

An example from 1792 illustrates early evidence of both
the need for quality assurance and quality assurance working
effectively to deliver to the client a product fit for its

purpose. William Jessop, then the engineer for the Cromford
Canal, had already lost his contractor and was continuing
with direct labour plus sub-contractors. Failure occurred
in the arches of the Amber Aqueduct at Bullbridge. In 1793,
on the same project, the Derwent Aqueduct near Cromford
failed by cracking 'for want of sufficient strength in the
front walls'. Jessop finally concluded that this was due to
his specificaton of a l.me which had not set.

So the need for quality assurance is not new and has
affected even the great works of our illustrious
predecessors. What is perhaps even more interesting is how
the quality assurance systems of the day protected the
client.

In response to the failure at Bullbridge, Jessop offered
to pay £650 from his own pocket and give up his salary both
retrospectively and until the canal was finished. His £650
was accepted but he was allowed to keep his salary. He
produced a remedial design for the Derwent Aqueduct
involving tie-bars and internal counterforts and he paid for
it himself. He wrote 'I think it common justice that no one
ought to suffer for the faults of another'. Perhaps salary
and fee levels in those days were such that one could truly
exercise a duty of care. However, the ethics which
motivated Jessop are still enshrined in the rules for
professional conduct of the Association of Consulting
Engineers. Phrases such as 'A member shall discharge his
duties with complete fidelity' and 'A member shall order his
conduct so as to uphold the dignity, standing and reputation
of the profession' still exist and if interpreted strictly
could lead to actions similar to those of Jessop - insurers
permitting.

The most valued asset a consultant has is his professional
reputation and any threat of its loss is a strong sanction
in the hands of a client. Thus most contracts between
professional advisers and clients define duties and
obligations and terms of payment. Generally these do not
address the issue of the designs being fit for their
purpose.

There is ample evidence that a significant number of
quality failures are generated in the design office. The
BRE report on low rise houses attributes 50% of its
catalogue of failures to design. In the related topic of
'buildability' where designs are accused of either being
impossible, difficult or costly to construct there are
examples of inadequate design. Among a group of research
projects into buildability undertaken at Loughborough one
studied the source of buildability problems. The identified
stage in the design process where many design details giving
difficulties originated was generally on the drawing board
of young, inexperienced engineers, and the most common cause
of this inadequate detailing was a lack of supervision. It
is to be hoped that the arrival of quality assured systems
is likely to provide an incentive to tighten up supervision

where it is needed. As fee competition and other pressures grow the need to have quality assured systems increases.

The other party in the construction process is the contractor and here a historical review reveals a different approach to client-contractor relationships. Early contracts were onerous. Clearly the contractor could not be trusted to discharge his duty of care with the same honour. The example quoted, from a contract at the end of the 19th century, reads as follows

> In the event of anything reasonably necessary or proper to the due and complete performance of the work (of which the Engineer shall be the sole judge) being omitted to be shown or described in the said drawings or schedules the contractor shall notwithstnding execute and provide all such omitted works and things as if they had been severally shown and described without any extra charge and according to the directions of the Engineer and to his satisfaction.

It is little wonder that such contract conditions let to the formation of the FCEC and the negotiation of more equitable contracts. But the roots of present contracts can be seen which require 'as can be foreseen by a reasonably experienced contractor' and 'to the satisfaction of the Engineer'. The existence of the independent Engineer and the inspection of the works by his representative on behalf of the client keeps firmly rooted in construction the separation of design and construction and the inspection mentality. How often has a contractor defended himself by the immortal words 'but the RE accepted it!'

The modern day requirements of quality assurance are described in BS5750 and the differences that exist between the requirements of BS5750 and current Conditions of Contract such as the ICE 5th Edition are substantially centred on the difference between inspecting for quality and planning for it and building it in. the main differences come under the general headings <u>Quality systems, Quality personnel and Materials, methods of work, inspection and testing</u>

Quality systems

BS5750 states how a supplier should assure quality through a quality system and requires that a quality system be established and periodically reviewed. The ICE 5th Edition asserts that the contractor shall adhere to the contract and hence to the quality standards set by the Engineer (e.g. clauses 36(1), 8(1)). This places responsibility on the contract documentation, especially the specification, and on the Engineer's judgement.

Quality personnel

BS5750 requires the supplier to appoint a managerial representative responsible for quality management,

preferably independent of other functions. It also requires
all personnel to have responsibilities and authorities with
respect to quality delegated and defined. BS5750 also
requires the supplier to identify training needs and
certification requirements.

ICE 5th requires supervision by competent persons and
requires the contractor's authorized representative or agent
to be approved by the Engineer (clauses 15(1),15(2)). The
reference to quality is implicit, not specific.

BS5750 offers the purchaser the option of appointing a
representative to obtain assurance. ICE 5th defines the
duties of the Engineer's representative, whose appointment
is not an option (clause 2(1)). Additionally the ICE 5th
gives the Engineer wide powers and, in this respect, is
probably stronger than BS5750.

Materials, methods of work, inspection and testing
BS5750 places responsibility on the supplier for

(a) inspecting, storage and handling of materials
(b) manufacturing methods under controlled conditions and
    for documentation that describes the manner of
    manufacture
(c) inspection, testing and furnishing of evidence of
    conformity with specification
(d) the development and maintenance of clear documented
    instructions of specified requirements
(e) the maintenance of records and the demonstration of
    achieved quality.

Against these awesome responsibilities placed on the
supplier by BS5750 under the ICE 5th Edition the role of the
Engineer is paramount and much of the quality management
measures are left to the Engineer to initiate.

(a) Materials must be to the approval of the Engineer.
(b) The Engineer has power to order removal and re-work.
(c) The Engineer has power to request information on
    contractors' methods.
(d) Work shall not be covered up without the Engineer's
    approval.
(e) The contractor shall provide for the tests required by
    the Engineer.
(f) The Engineer keeps records of as-built work and
    quality checking.

Clauses 39(1),14(3),36(1),38(1) are examples. Therefore,
with respect to quality management the key differences are
that the contractor is often, apparently, responding to
requests by the Engineer. The Engineer has much wider
powers under ICE 5th than the purchaser's representative
under BS5750 but with these wide powers is a monumental
responsibility. The ethos in construction is clearly one of

9

inspection.

However, well managed contractors have already gone a long way down the road of quality assurance without glorifying their practices with that title. Most contractors

(a) have a materials quality system in place to some degree
(b) buy from known suppliers
(c) employ only experienced and qualified staff
(d) produce method statements for their own internal instruction and use
(e) keep amazingly detailed records
(f) can, in most cases, demonstrate quality is achieved.

Thus if clients add quality assurance clauses to their contracts, most reasonably sized contractors will respond without excessive difficulties because in the end quality assurance is the current buzz word for good management practice and good management practice in the end is the most economic form of management practice.

What of the future? QA is inevitable. More clients will add quality assurance clauses to their contracts. As well as the addition by clients of QA clauses to existing forms of contract new forms of contract will emerge.

The British Property Federation's form of contract took a small step towards a redefinition of the relationship between design and construction by allowing the contractor to undertake some detailed design. This was not the final step down this road and the exploration of this avenue in other forms of contract is already under way.

Contractors will continue to develop greater self-reliance and grow away from the inspection mentality. As the concepts of quality assurance gain hold contractors will feel the need for greater management control of the whole construction process. There will be more design and construct contracts and more design firms bought up by contractors.

On the engineer's side, specifications will develop greater precision and the need to specify requirements in measurable standards will grow. The Resident Engineer will become the purchasers' representative and will probably also retain many of his existing powers with, it is to be hoped, a reducing need to exercise them.

Those in research and teaching will continue to have a useful role to play in the development of standards, means of measuring them and in developing methods and procedures.

The move to QA is not a tinkering with the construction industry. It is a fundamental shift in basic approach. There will be difficulties, but one overriding fact remains - the client wants quality. The construction industry has no choice but to deliver it.

1. Quality assurance in the design office — is it cost effective?

J. M. DUNCAN, BSc, MICE, MIQA, Design Group Partnership

SYNOPSIS. All Civil Engineering related Businesses, be they production, construction, service or consultancy, are budget controlled or profit motivated. Quality Assurance is a Management Control System purporting to improve communication, reduce error and hence ultimately improve the ability of a Company or Service group to compete in the marketplace, increase profits or maintain budgetary control. If Quality Assurance cannot be perceived to provide such benefit then its application will be ineffective, painful and ultimately damaging to the Company or Group concerned.

The following examines the experience of one Consulting Practice, Design Group Partnership, who adopted a QA system in May 1985 and were granted a registration by BSI to BS5750 part 1 in May, 1987.

INTRODUCTION

1. There is no doubt that the introduction of a formal quality assurance system represents real additional costs in excess of the overheads previously experienced by the company. Its application can only be perceived to benefit a company if this can be realistically balanced against the costs of reduced errors and increased efficiency. In the following paper I have attempted to clinically address both these aspects from the experience of operating a formal quality assurance system for over three years.

2. It is worth emphasising at this point, that the costs examined in this paper do not include certification and registration of a quality assurance system with an external third party assessor. Such certification does little to enhance the ability of a company to implement a quality assurance system and the costs therefore, should be bourne directly on a Public Relations budget or against a cost centre associated with a client group demanding Q.A. certification as a pre-requisite to consideration for commission.

THE QA OVERHEAD

The costs of operating a formal QA system in a design office could be considered in five distinct areas:-

1. System Development: The real cost of system development will depend greatly on the degree to which formalised procedures already exist within the company. Most consulting groups operate formal procedures to some degree and hence the notion that QA is novel is misleading. The development of the system must be carried out by senior managers who have the total support of the Company Directors and a thorough working knowledge of the company practices. If the individuals concerned are trained and motivated correctly, an average budget to 500 manhours would be reasonable for a Company whose working procedures are logical and reasonably well developed. An allowance of 120 hours for training senior managers in systems development should also be considered.

2. System Implementation: The success of any system is entirely related to the degree to which the principal staff are trained. During the implementation phase, sufficient time should be given to the training of staff in use of the system and guidance given on a regular basis on its application to the working environment. An allowance of 2 hours per person should be allowed for initial training and an additional one hour per person for on the job support from a source external to the project team.

3. System Maintenance: Items 1 and 2 above are one-off expenses, however, system maintenance and item 4 below, training, are continuing budget items. System maintenance involves the process of carrying out audits on the system and the projects within the company, adjusting the system to suit changing markets and internal requirements and maintaining records of all quality related activities within the company. The costs of system maintenance are combined with item 4 below.

4. Training: The importance of training new members of staff in the company procedures cannot be over emphasised. Taking 3 and 4 together the requirement for a QA Manager either part time or full time is obvious. In smaller businesses the duties of a QA Manager could be combined with other administrative duties such that the total overhead is mitigated. Alternatively, services are now being offered by various agencies to provide "Q share", whereby the services of a QA Manager are shared between a number of firms, thus again reducing costs and overhead dramatically. A reasonable allowance would be 5 manhours per annum per person employed for the services of a Quality Manager. The above figures are derived from the experience of a Company employing 200 technical and professional staff operating a multi disciplinary Engineering Consultancy. An allowance of 2 manhours per employee should be allowed per annum for attending training lectures.

5. System Operation: The development of project plans and participation in audits are two activities which would not be essential without formal Q.A.   It is difficult to budget for project plan development since, with a 'streamlined' system, it should not have any great impact on the overall project manhours.   A nominal total of 250 hours per annum of Senior Management time could be allocated for this activity.   Each individual within the Company will participate in audits for an average of 2 hours per annum.

Budget Summary for a Company Employing 200 Staff (total)

A.   Initial Outlay:-

Senior Management Training in system development:	120 hours @ £20 =	£ 2,400
System Development:	500 hours @ £20 =	£10,000
System Implementation:	3 x 200 @ £20 =	£12,000
Secretarial Support		£ 2,500
Documentation (hard copy costs)		£ 250

£27,150

B.   Annual Maintenance, Training and System Operation:-

QA Manager or 'Q Share'	5 x 200 @ £20 =	£20,000
Training (average £12 per hour)	2 x 200 @ £12 =	£ 4,800
Project Plans	250 @ £20 =	£ 5,000
Audits	2 x 200 @ £12 =	£ 4,800
Secretarial Support		£ 2,000

£36,600

If we assume, therefore, that 170 members of staff are involved in 'productive' activities, then the total reimbursible hours worked per annum would be: 170 x 1800 = 306000 hours.

13

The increase in overhead for maintenance, training and system operation would be $\frac{£36,600}{306000}$ = £0.12 per man hour

NB: The above represents only those sums considered to be additional to the overheads experienced by any reasonably well organised Consulting Practice. Item 'A' costs are an initial investment and should be incorporated into the company accounts in accordance with the individual corporate policy.

THE CONSEQUENCES OF ERROR

1.   Many documents promoting the benefits of Quality Assurance refer to statistical evidence related to the construction industry incidence and source of failure / error. The experts differ in their interpretation of this evidence, however, it is generally agreed that in terms of cost penalty to the overall project, at least 30% of the blame lies with the design office function. It would appear that the researchers have concluded that the cost penalty is realised due to:-

a)   Inadequate or incorrect specification at tender.

b)   Inadequate training and management of the designers responsible for producing calculations and drawings. This is realised in excessive changes to the detail information throughout the construction period of the project and consequently variations to the contractors costs.

c)   Poor communication between the principal parties in contract, which ultimately leads to confusion and cost related delays.

d)   Inadequate definition of responsibility within the Design Management Group.

2.   Any student of Quality Assurance would immediately identify the above as the major factors influencing the necessity for development of QA with the design office.

3.   However, in addition to the above, the Design Group also often suffer a substantial cost penalty due to the continual recycling of information which does not necessarily have an impact on the construction costs.

4.   We must accept that no management system such as Quality Assurance is ever applied totally and therefore, its effect would, at best, only partially eradicate the problems detailed above.   In addition, the problems associated with recycling information can be equally instigated by any of the parties involved and hence, if only one party adopts Quality Assurance as a management system, even if totally applied, its effect can never be fully realised.

THE EFFECT OF QA

1.   Since Quality Assurance is deemed to be a total management system, applied throughout the company, its effectiveness is best related to the overall performance of the company.   In order to make an assessment of performance, the following questions should be asked:-

a)   Do the clients perceive an improvement in performance?

b)   Are the senior managers within the company more confident of company performance following the introduction of Quality Assurance?

c)   Are the individual projects better able to meet fee limits and budget control?

2.   With regard to client perception of the effect of Quality Assurance, the argument can only be presented in terms of the overall cost and efficiency of the project.   In terms of fees paid to the consultant or designer, the client can only expect the increased internal efficiency to off-set the additional costs of maintenance and training.   For all clients 'Time is Money' and the ability of a formal QA system to improve the time efficiency of a project should not be difficult to argue. It is the experience of Design Group Partnership that there are two major effects of Quality Assurance perceived by our current clientele:-

    The information supplied at tender is far more comprehensive and accurate which ultimately leads to a far greater confidence in the ultimate budget constraints being met.

    The enhanced quality of the information supplied at tender and the structured communication between all parties throughout the period of project leads to a much greater confidence that the works will be completed within the original programme period.

3.   With regard to the effect within the company of applying a formal QA system, there is no doubt that the individual managers benefit greatly in the long term.   Far from being the burden as some opponents of formal QA would have us believe, a well applied system provides the manager with the opportunity to delegate the routine functions with confidence and therefore concentrate on the development of engineering and problem solving which are the essence of his existence.

4.   In addition, the formalisation of design review and the development of comprehensive Quality Plans allow the total project team to become more involved in the Engineering and Management function.   Both of these aspects of formal QA are excellent vehicles for training junior members of the team and ultimately serve to enhance confidence throughout the Company.

5.   Ultimately, the success or failure of a QA policy will be judged against the ability of a company to meet fee limits and budget control, which will be reflected in overall company performance.   It will take some time before the system is "second nature" within the Company and therefore premature assessments may be misleading. However, it is the experience of Design Group Partnership that, on those projects where the system has been totally applied, the effect has been to dramatically improve our ability to meet such targets. It is our expectation and experience that in such cases the fee limit or budget will be met with considerable accuracy.

CONCLUSIONS.

1.   Referring back to the beginning of this paper, the question to be answered is "Does quality assurance improve the ability of a company or service group to compete in the market place, increase profits or maintain budgetary control". The answer to the question is yes, with the proviso that the system is applied from the most senior level of management through to the junior trainees in a comprehensive and enthusiastic manner. However, if it is applied in a cosmetic manner, then it will have a negative effect with respect to these objectives. Likewise it follows, that for the system to be effective, it must be largely an expression of the manner in which the company already operates and wishes to operate and not a reflection of a set of rules dictated by an external third party. Any guidance documents or standards used to develop the system should therefore, be used only to give a basic framework in which the companies desired management system can be emplaced.

The confidence with which I make the above statement is based on the following simple principles:-

    (a)   The majority of error that occurs within the
           design function is associated with inadequate
           supervision, training and communication within the
           management team. Quality assurance is a
           management system designed to tackle these
           problems and therefore by definition, a well
           applied system will have a sub-stantial effect in
           eradicating them.

(b)   No business can exist without clients and the best client is the one who offers repeat business. The most marked effect of quality assurance, in the experience of Design Group Partnership, is the ability to maintain client confidence. It therefore follows, that an effectively applied system will substantially increase the ability of a company to maintain business and expand its client base.

(c)   There can be little doubt that quality assurance is here to stay. Many of the UK based major client bodies now perceive a time in the very near future when the award of commission will be based on the consultants ability to demonstrate quality assurance in addition to traditional selection procedures. It will therefore become impossible within a relatively short period of time to compete for major projects without a visible quality management system.

(d)   The real cost of implementing the system should not be underestimated, however in our experience it is relatively insignificant in comparison with the overall benefit accrued by the client. It is certainly true that the increased overhead associated with the formal system will eat into profit margins, however, this is more than matched by the effects of increased efficiency experienced as a result.

2.   Referring briefly to the role of central government in promoting Quality Management and the continuing trend towards fee competition,it is perhaps worth noting that, in the immediate future, these objectives may well be incompatible. The introduction of formal Q.A. does place a real cost penalty on the consulting practice and the benefits accrued in the long term are mainly experienced by the clients. It is difficult therefore, for the consultant making this substantial investment, on behalf of his clients, to compete for design commissions with those who have not made a similar investment. In the long term, when a significant number of firms have recognised Quality Management Systems, it may be appropriate to apply fee competition, however, this will not be the case for some time to come. Likewise, as stated above in (d) the ultimate effect, once the system has become well established, will be that consultants are better able to meet fee limits and budget constraints. It would be a great injustice, therefore, for firms with a well established track record to be penalised in fee competition, whilst taking measures which will ultimately be to the greater benefit of the client.

17

3. When assessing the effect of Q.A. in enabling a business to compete, the emergence of the European Market in 1992 should not be ignored. Clearly, we have a major opportunity to establish, on a Nationwide basis, our intent to provide top quality services. The costs associated with the development of Q.A. could be greatly eased by the production of model documents, which are currently being drafted. The costs of certification might also be mitigated if we, as a profession, take a more forward role in the establishment of the ground rules and direct the certifying bodies towards the requirements of consulting practices.

4. Finally, if it can be accepted that quality assurance is both desirable and essential it is perhaps worthwhile re-emphasising that no management system will ever be effective and beneficial unless all those associated with its development and implementation are totally convinced that it will ultimately benefit the company. In addition, as stated previously, the system will fail from time to time and those failures will be more painful because of the expectations encouraged by the development of the system. The senior managers must therefore be prepared to accept the disappointments and maintain their enthusiasm to "get it right first time, next time".

2. The application of quality systems to civil engineering construction

K. A. L. JOHNSON, BSc(Eng), FICE, Fairclough Civil Engineering Ltd

SYNOPSIS
There has been, and continues to be, much debate over the relevance of Quality Systems in the Construction Industry. However, more and more clients are specifying these requirements. Designers and contractors have had to respond by developing systems. This paper outlines and discusses the development, implementation and costs of Quality Systems (in accordance with BS 5750) for contractors.

EXISTING PRACTICE

1   The contractor has always been responsible for planning, methods, plant, labour, and ensuring that the specified requirements are met.

2   It is normal to establish a site management structure with an Agent as the contractors' principal representative and to define the division of responsibilities between head office and site.

3   Systems vary between organisations, but in any organisation the system needs to be able to collect information so that decisions can be made on the best information. Decisions then need to be communicated so that the necessary actions can be taken.

4   It has always been the intent to use methods of carrying out the work to avoid errors, to keep accurate records and give people the freedom to act. Therefore, the construction industry has for many years applied management systems very similar to most of the requirements of BS 5750.

5   However, some of these systems are not comprehensive and the application can be patchy; some activities are well executed while others may not be fully covered.

6   In any management system prevention must be the priority. It is more efficient to prevent problems than to spend time and money solving them.

7   Recent studies by the World Bank have suggested that most project faults result from the failure of management systems.

8   Other studies by the Building Research Establishment have suggested that problems arise largely through management short falls in design and the communication of design into construction.

9   Every day brings action from committed, dedicated,
people in our industry who want to do things in their
traditional manner. This of course, is a major obstacle
to progress. People must recognise that improvement is
necessary before any progress can be made.
BS 5750   QUALITY SYSTEMS

10 Currently most contracts require the application of a
Quality System based on BS 5750 (ref 1). This British
Standard was originally published in 1979, updated and re-
issued in June 1987 and is now identical to ISO 9000 (ref 2).

11 BS 5750 was originally written for the manufacturing
industry and therefore some of the terminology is foreign
to the construction industry. However, with the amended
standard, this gap has been partly bridged, although some
difficulties with terminology remain.

12 The important point is that the principles are basically
sound and set out an overall requirement for Quality
Management. It is the application and implementation
that needs attention.

13 Before discussing the application of BS 5750, it is
necessary to summarise some of the basic definitions. The
definitions are from BS 4778, part I 1987 (ref 3):

Quality   The totality of features and characteristics of a
product or service that bear on its ability to satisfy stated
or implied needs.

Quality Policy   The overall quality intentions and direction
of an organisation as regards quality, as formally expressed
by top management.

Quality Management   That aspect of the overall management
function that determines and implements the quality policy.

Quality Assurance   All those planned and systematic actions
necessary to provide adequate confidence that a product or
service will satisfy given requirements for quality.

Quality Control   The operational techniques and activities
that are used to fulfil requirements for quality.

Quality System   The organisational structure, responsibilities,
procedures, processes and resources for implementing quality
management.

14 'Quality' is not used to express degree of excellence in
a comparative sense, nor is it used in a quantitative sense
for technical evaluation.

15 The attainment of the specified quality requires the
commitment and participation of all members of the organisation,
whereas the responsibility for quality management belongs
to top management.

16 Within an organisation, Quality Assurance is a management
tool and serves to provide confidence in a supplier.

17 BS 5750 clause 4, gives the quality system requirements,
as follows:
1   Management Responsibility Policy, Organisation, Review.
2   Quality System preparation, implementation.
3   Contract Review.

4   Design control responsibility, inputs and outputs, verification, changes.
5   Document Control.
6   Purchasing, sub-contracts, data for products, verification.
7   Purchaser supplied product.
8   Product identification and traceability.
9   Process Control (Construction control), plan, method statements, monitoring and control.
10 Inspection and testing.
11 Inspection measuring and test equipment.
12 Inspection and test status.
13 Control of nonconforming product.
14 Corrective Action.
15 Handling, storage packing and delivery.
16 Quality Records.
17 Internal Quality Audits.
18 Training.
19 Servicing.
20 Statistical Techniques.

18   Before describing a system for contractors, it is important to decide the boundaries within which it is to apply. One must examine the various parties to the construction project: client, designer, contractor, sub-contractor, suppliers and the operator. Also to consider what was done in the past, what is being done presently, and how the situation may develop in the future.

19   To obtain the best result for a project it is necessary to apply Quality Systems from the beginning. The client should have a Quality System to control his brief and define his requirements. The designer needs a Quality System to control the design process and to ensure that the clients brief is satisfied. Next, the contractor needs to have a Quality System which can satisfy the requirements of the designer and execute the project in accordance with the specified details. Finally, the operator must operate in accordance with a Quality System to ensure the structure or plant is not misused. Only with this overlapping of Quality Management will the full benefit be derived.

DEVELOPMENT
20   I do not intend to comment on, or go through one by one, the twenty quality system requirements of the British Standard. Anyone who takes the trouble to examine them will see that they are all necessary and relevant to the management of any business. Individual organisations must decide how best to apply them in their particular case.

21   Quality Systems are concerned with conformance to specified requirements and that activities are carried out in accordance with written instructions. The Quality System should only be as comprehensive as is needed to meet quality objectives.

22   To produce a documented quality management system, it

is best to work from existing company management methods and technical practices.

23 It is important to note that it is quite possible to operate non QA contracts from a QA managed company, but it is extremely difficult, to operate QA contracts from a non QA managed company.

24 In the development of a quality system it is essential that top management take the lead, and that they involve all levels of staff within the company.

Company Quality System

25 The first priority is the production of a company quality system. This consists of a manual, containing a quality policy, comprising statements on quality requirements, authority and responsibilities, organisational chart, and method of distribution, amendment and re-issue of the manual.

26 Also the manual contains procedure outlines for the control of the business; an index of detailed procedures, which are issued and authorised for use and a list of registered holders.

27 Once a company quality system has been developed it is then very simple to develop particular quality systems for individual contracts which are defined as quality plans.

Contract Quality System

28 The quality system for each contract will comprise a manual, setting out policy, responsibilities, organisation structure, and procedures, all derived from the company system. Any requirements that are not common to the company system will need to be covered by procedures which set out the manner in which a specific contract operation is undertaken. Gradually a library of work procedures will be built up, which can be applied to standard operations in any contract.

QA Department

29 A company quality system requires a QA department. This department will administer and control the development of the manual and the amendment and re-issue as necessary. The QA department will assist in preparing and in the implementation of procedures and carry out audits.

30 It is important to define the role and authority of the QA department and its relationship with other service and operation departments.

Head Office Systems

31 The company quality system will define the requirements for each department to establish its own operational procedures (eg buying, design).

Procedures

32 Procedures define the manner in-which control is exerted over various operations. It is essential that procedures are written by the people who will use them and reflect the manner

in which company activities are carried out.

33 Procedures should always be kept simple and not incorporate unnecessary constraints. A flow diagram can be helpful in some cases. Procedures should be no more than sound commercial sense committed to paper. There are basically three types of procedure. Administration, quality, and works.

34 The appendix includes an example of a quality procedure.

Sub-Contractors and Suppliers

35 Logically each supplier should have his own quality system. Products are probably the easiest area to understand and identify with because for many years there have been products to British Standards identified by the BS Kitemark.

36 Sub-contractors should be encouraged to develop their own quality systems that suit their particular existing systems of management and control. If the subcontractor does not have a quality system there is a tendency to force him into the main contractor's system or that of the client. This is a certain way of leading to confusion and inefficiency.

IMPLEMENTATION

37 Having developed a quality management system it is then necessary to implement it. This is probably the most difficult phase. The first essential is leadership. This can only come from top management, without their commitment and belief implementation will be ineffective.

38 Also middle management must be fully educated in the needs and benefits of the system relative to their particular areas of responsibility.

Attitudes

39 The operational level must be clear on the objectives. My experience is that this level is the most receptive once the system is explained; because they have an opportunity to be involved in and contribute to the development of procedures which determine what they actually do. Significant benefits in communication and understanding result.

40 Reaction from middle management is much more diverse. This group has a different set of problems. Many consider that existing systems are basically sound and therefore they should not change. Also the industry suffers from the 'not invented here' syndrome. A major difficulty is that the construction industry has a number of 'macho people', who like to manage by the seat of their pants, with the excitement this gives.

41 Top management in many organisations do seem to be sitting on the fence. They are not sure whether QA is necessary or whether they can afford 'it'.

Training

42 Apart from establishing specilist QA Staff, it is necessary to train and educate all staff at all levels.

⌗ Fairclough

Fairclough Civil Engineering Limited

QUALITY PROCEDURE		SERIAL NO. QP	0 1 2 / 0 0 1

ENG/ORDER NO. ☐☐☐☐☐☐☐

SHEET ☐1 OF ☐2

CONTRACT NO. ☐☐☐☐☐☐

PROJECT ☐

CONTRACTOR/SUB-CONTRACTOR | FAIRCLOUGH CIVIL ENGINEERING LIMITED

TYPE OF WORK   SUPPLY AND PLACEMENT OF CONCRETE

TITLE

Documents Required: (Prior to Commencement)	Document Reference	Issue No.
1. Quality Procedure Sheets 1 and 2.	012/001	
2. Works Procedure Sheets 1 - 5 inclusive.		
3. Approval of manufacturers and suppliers and the sources materials listed:	012/WP/001	
READY MIX CONCRETE, CURING AGENTS, REINFORCEMENT, WIRE TIES, WATER BARS, JOINT FILLERS AND SEALANTS, HOLDING DOWN BOLTS, NUTS AND WASHERS, TESTING HOUSE	FS1	
4. Drawings		
5. Manufacturers and suppliers certificates, together with additional information to confirm products listed above comply with specification and drawings.		
6. Laboratory test reports on initial samples of cement, fine and coarse aggregates, water and reinforcement.		
7. Method for batching, mixing and transporting concrete including the appropriate calibration certificates.		
8. CONCRETE MIX DESIGN.		
9. Laboratory test results to show compliance with specification for strength and workability.		

Signed for the Sub-Contractor* .... Date ....
(* if applicable)

Signed for the Main Contractor .... Date ....

Approval is granted and the Contractor is clear to proceed with the work covered by this Quality Procedure

Signed for RE/SO .... Date ....

DISTRIBUTION BY
1. Original to be retained by Client
2. Photocopy to be returned to the Contractor.   * Delete as applicable

Fig.1. Company quality procedure documentation

24

SHEET 2 OF 2	SERIAL NO. QP 0 1 2 / 0 0 1		
Documents/Inspection Requirements as Work Proceeds	Hold Point Approval of RE/SO	Routine Submission to RE/SO	Specific Submission to RE/SO
1. Results of regular sampling of testing (or certificates) of the following materials listed; cement, water fine and coarse aggregates and reinforcement in accordance with the requirements of the Specification and Drawings.		X	X
2. Notification and approval of changes for material listed in 3 overleaf.	X		X
3. Results of concrete testing in accordance with the requirements of the Specification.		X	X
4. Record of locations in the structure represented by the samples taken.		X	X
5. Checks on dimensional accuracy of encast items.		X	X
6. Completed and approved pre-concreting notice and post-concreting inspection sheets FS2 and FS3.		X	X
7. Completed and approved defect notice and remedial work inspection form FS4.	X		X
8. Concession applications raised in the course of contract, FS5.		X	X
9. Technical query forms raised in the course of the contract, FS6.	X		X
10. Record drawing showing pour numbers sequence of pouring, date of pour.		X	X
11. Foundation check and acceptance sheet FS10.		X	X

All Documents and Records required by this Quality Procedure have been submitted and the stipulated instructions carried out in accordance with specification and agreed deviations.

Signed for Sub-Contractor*
(*if applicable) .... Date ....

Signed for Main Contractor
.... Date ....

All documents present and verified
Signed for RE/SO
.... Date ....

Fig.1. - continued

43 Training falls into three general areas. Firstly, there is awareness of the concepts and principles of quality assurance. Secondly, an introduction and explanation of the quality system itself and finally, nuts and bolts training in the application of the system where it specifically applies in the individuals' area of operation.

44 Writers have identified five stages in the implementation of quality systems, starting with uncertainty, moving on to awakening, eventually turning into enlightenment, then knowledge and finally certainty in the actual application of the system. This is a good yardstick to measure progress in the implementation of the quality system at the three levels already discussed; operational, middle management, and top management.

Audits

45 Audits can achieve four main things: identify the root cause of any problem, make constructive criticisms for compliance or improvement, evaluate any action necessary and act as a catalyst for improvement and conformity.

46 The result of any audit should either produce a corrective action for conformance with the system (which is established sound, efficient and logical) or identify that the system is inadequate and should be modified. Hence it is important to recognise that quality management systems are live systems and not tablets of stone.

47 A continuous problem with any documented system is the essential requirement to keep it updated. The actual activity of any management system must correspond with the documentated requirement.

Accreditation

48 For products it is quite straight forward and third party accreditation is the appropriate answer. Also, for some types of specialist subcontractor third party accreditation may be equally applicable. For a number of major clients second party assessment is a very logical and sensible route. There continues to be the doubts over third party accreditation for both main contractors and consultants particularly over the criteria against which it should be carried out.

49 The problem with second party assessment is that many companies could be subjected to multiple assessments. The full extent of this difficulty is not easily quantified.

50 If a company has its own effective and efficient QA department, first party assessment will result naturally. This will go a long way to assist in obtaining second party assessment and third party accreditation.

COSTS

51 There are costs in producing a quality system and applying it. These can be covered under a number of headings. The initial preparation of a company quality system; employing specialist QA Staff; educating and training of all the Staff;

management time in the actual application on each contract and in head office.

52 Only on large contracts would full time quality assurance staff be employed. Normally the section engineer would maintain the quality records and procedures would be produced by the particular section carrying out that work.

53 The cost most difficult to quantify is those of implementation. This depends on attitudes and leadership. The cost of accreditation depends on the extent and type required.

54 Unless the system is preventing errors occuring, and reducing the number of failures, then the system is not correct. The cost of the system should be more than recovered by the prevention of mistakes, but this is not easy to measure.

55 If the system becomes too elaborate then the cost could exceed the gains. It is a balancing act to ensure value for money. This judgement will only come with experience in any particular area of the industry.

SUMMARY

56 It is essential that any quality system is both flexible and live. It must also be cost effective otherwise there is no point in having it.

57 The requirement for management to control operations and to learn from feedback is self-evident. Quality systems to the requirements of BS 5750, can provide a means to achieve these objectives.

58 The level of paperwork for documentation and traceability are all within our control. It is only by addressing these objectively rather that negatively will we arrive at the most appropriate level of paperwork for any particular project in the industry.

59 Any system developed must be a real system and not a copy of something somebody else has produced. It must be a system for the particular company and the specific contract. The quality system must be the system for managing the business and contract, it cannot be an add-on.

CONCLUDING REMARKS

60 With the appropriate system sensibly applied there are a number of benefits which can accrue: fewer failures, more precise specifications, a better and more structured form of communication, a full record for each contract and a sounder basis on which to select main contractors, suppliers and sub-contractors.

61 These benefits will only be obtained in full with the complete participation of client, designer, engineer, contractor, subcontractor and supplier.

62 As an industry we are overdue in getting our act together in preparation for the European Community.

Acknowledgements

The assistance of a number of colleagues in preparing the paper is acknowledged with thanks. The views expressed in the paper are those of the author and do not necessarily reflect those of Fairclough Civil Engineering Limited.

REFERENCES
1   BS 5750 - 1987 Quality Systems Parts 0 to 4.
2   ISO 9000 - 1987 Quality Systems   Four Parts.
3   BS 4778 - 1987 Quality Vocabulary.

3. US Army Corps of Engineers' system for quality assurance and contractor quality control

D. LAWRENCE, US Army Construction Engineering Research Laboratory

SYNOPSIS. There are many events before actual construction starts and after actual construction ends that affect the quality of a project. In the late 1960's, the U.S. Army Corps of Engineers developed a Quality Assurance/Contractor Quality Control system (QA/CQC). The QA/CQC system encompasses the construction process from the initiation of the project, the verbalization of what type of construction project is desired, to the opposite end of the scale, when the user has had possession for sometime. The Corps' CQC/QA system relies on every faction involved in building a project communicating and working cohesively to build a quality project which meets money, time and functional criteria set forth. Although the principles of the CQC/QA system are functioning mostly only on Corps' projects, through the efforts of ASCE and other organizations QC concepts are receiving more recognition for their capabilities and payback benefits.

1. There is a growing acknowledgment in the United States that to construct a quality project cost effectively, all of the factions from the owner to the designer to the constructor to the user must communicate and work together. Without this communication even though the contractor may construct the project to the designed criteria it may not be what the owner wanted. Therefore, in the owners eyes a quality project was not constructed, even though, to many a quality project is defined as a project built in conformance with the contract specifications and drawings.

2. The first step in producing a quality project is defining what the owner wants. The owner's expectations must then be matched with the user's needs and translated by the designer into concise requirements of what is to be built, how, with what materials and most importantly to what specifications. The contractor in coordination with the owner, designer and user must then translate the designed project into a quality project.

3. The U.S. Army Corps of Engineers wears many hats during the construction of their projects. Depending on the

project, the U.S. Army Corps of Engineers may be the designer and the contract manager as well as the end user or their role may only be one of contract management where the design was performed by an architect-engineer firm and the user may be the military. No matter what the project, the role of the U.S. Army Corps of Engineers is, as a minimum, to coordinate the needs of the owner, end user, designer and constructor to produce a quality project.

4. In the late 1960's, the U.S. Army Corps of Engineers developed a Quality Assurance (QA)/Contractor Quality Control (CQC) process. This QA/CQC process replaced a process which previously relied on the U.S. Army Corps of Engineers providing and performing quality tests to ensure that the construction was being built to Corps standards and specifications. The QA/CQC process clarified and therefore shifted responsibility for constructing a quality project back onto the shoulders of the contractor. Although the U.S. Army Corps of Engineers performs quality assurance, the contractor's responsibility for constructing the project to the specifications which are defined in his contract and performing quality control are not diminished.

5. Contractor Quality Control, in the U. S. Army Corps of Engineer's QA/CQC process, is defined as the construction contractor's system to manage and control and document his own, his supplier's, and his subcontractor's activities to comply with contract requirements. Quality Assurance is defined as the procedures by which the Government fulfills its responsibility to be certain the CQC is functioning and the specified end product is realized.

6. The U.S. Army Corps of Engineers Quality Assurance/Quality Control (QA/QC) process does not concentrate only on the construction phase, but encompasses the full spectrum of activities which impact quality. The activities range from the project conception and design phase through turnover and operation phase. Some of these activities are listed in Table 1. The sequence in which the major activities take place is depicted in Figure 1.

7. There are many pre-award activities that contribute to building a quality project, including identification of the owner's needs. The U.S. Army Corps of Engineer's performs a minimum of two specific design reviews. The first review, a biddability, constructability and operability (BCO) review, is performed when the design is 35% complete. At this point, the user/owner and the field entity responsible for QA during construction are encouraged to review the package. Reviewer comments are then incorporated into the design. When the design is basically complete, a final design review is performed by the engineering section of the Corps. Engineering's review comments are incorporated into the design and a final BCO review is performed by the

construction field entity responsible for QA and the user/owner. Their comments are once again incorporated and the contract package is advertised for bid. A pre-award survey is then made on the responding contractors. At this point, the knowledge gained from the contractors past performance becomes important. The Government awards the contract to the lowest responsible bidder. The contractor's past performance records are thoroughly collected and evaluated. Using objective criteria, the contractor's past experience on projects at the technical level of the advertised projects, his financial ability to perform the contract, his ability to comply with time requirements or delivery dates (taking into consideration existing building commitments, etc.) a responsible and capable contractor can be chosen and awarded the contract.

8. Countless problems can be avoided at a very low cost at the design review stage of the project. At this stage, the guide specifications should be reviewed to assure that they establish the desired technical provisions, establish the specific requirements desired and are current with technology. The initial BCO review should include a check of the physical layout, determination of accessibility for construction equipment and determination of utility availability. By properly performing these reviews and incorporating the results, numerous modifications in cost and time can be avoided.

9. The next major group of activities involves the planning of quality assurance. For the U.S. Army Corps of Engineers, some of these activities include inspecting the site, reviewing utilities and site specifics, developing organizational plans, determining test requirements, determining corrective action plans, staffing requirements and recruiting, etc. By planning ahead, the QA staff has time to supplement their capabilities, if necessary, by attending Corps-sponsored training classes. Basically, the QA planning allows the Corps to obtain a properly staffed and trained QA group in place ready to perform in unison when construction begins.

10. Prior to the start of construction there are two meetings between the contractor and Government that are contractually required. One of the meetings, the preconstruction conference, covers all aspects of the contract. Specific responsibilities and authorities are detailed and agreed upon at this meeting. In general, contract provisions, personnel requirements and procedural matters for administering the contract are established.

11. The other meeting is the QC/QA mutual understanding conference. The purposes of this meeting are: (1) to achieve a mutual understanding with the contractor of his role in quality control, (2) to review the QC Plan with the

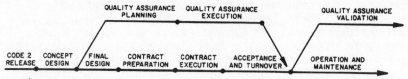

Fig.1. Quality assurance/quality control process model

PROJECT CONCEPT AND DESIGN PHASE	PRECONSTRUCTION PHASE	CONSTRUCTION PHASE	TURNOVER AND OPERATION PHASE
PROJECT DESIGN DESIGN APPROVAL CBO REVIEWS STAFFING PLAN	PRE-CON CONFERENCE CQC PLAN GOVERNMENT ESTIMATE REQUEST FOR PROPOSAL QA PLAN TIME ESTIMATE CONTRACT AWARD	QUALITY CONTROL EXECUTION QUALITY ASSURANCE EXECUTION CONTRACT ADMINISTRATION TESTING SUBMITTAL APPROVALS CONTRACTOR PAYMENTS CPM SCHEDULING 3 PHASE INSPECTION TITLE II SERVICES DAILY LOGS/REPORTS MODIFICATIONS CLAIMS SAFETY MANAGEMENT	FINAL INSPECTION ACCEPTANCE TURNOVER CONTRACTOR RATING POST COMP INSPECTION WARRANTY INSPECTION DEFICIENCY CORRECTION FISCAL CLOSEOUT

Table 1. Areas impacting quality

contractor (the government must receive, review and accept the QC Plan before the meeting) (The purpose and contents of the QC Plan are detailed later), (3) to establish a good working relationship between the Government and the contractor. At the conclusion of these two conferences, there should be a mutual understanding of what is expected of the contractor and that the achievement of building a quality project is a shared goal of the contractor and the Government.

12. During construction the Corps' QA staff has several tasks to perform. Among these tasks are the following: (1) Require revision to the CQC plan and its execution when necessary, (2) verify adequacy and calibration of test equipment, (3) spot check CQC approved submittals, (4) perform periodic jobsite assurance conferences on QA/QC, (5) participate in the three-phase inspection process as appropriate, (6) conduct government QA tests at the jobsite, (7) monitor corrective actions for promptness and control, (8) insure that no work is placed on unapproved or unaccepted work. The performance of these tasks is intended to ascertain that the CQC is functioning and the specified end product is realized.

13. The contractor is required to submit and receive approval on a quality control plan prior to beginning construction. The QC Plan describes how the contractor proposes to implement his requirements for inspecting the construction. The plan includes: (1) a description of the

QC organization with a chart showing lines of authority.
(The QC staff is required to report to a project manager or
someone higher in the contractor's organization), (2) the
name, qualifications, duties, responsibilities and
authorities of each person assigned a QC function, (3) a
copy of the letter to the QC manager signed by an authorized
official of the firm, which describes the responsibilities
and delegates the authorities of the QC manager, (4) the
general approach to the methods of inspection (the three
phase inspection concept, and procedures for scheduling and
managing submittals, including those of subcontractors,
offsite fabricators, suppliers and purchasing agents), (5)
control testing procedures for each specific test, (6)
reporting procedures including proposed reporting formats,
(7) A list of the definable features of work, (A definable
feature of work is a task which is separate and distinct
from other tasks and has separate control requirements.).
If the QA staff determines the CQC plan is not adequate once
construction starts, the contractor is required to modify
the plan. This plan is intended to provide the CQC staff
with guidance to perform their QC duties.

14. The CQC staff is responsible for insuring the
contractor complies with the contract plans and
specifications. Some of the QC's duties are: (1) shop
drawing approval, (2) submittal review and approval, (3)
deficiency identification and tracking, (4) three-phase
inspection, (5) documentation of construction progress and
testing performed, (6) safety, (7) QC testing. Most of
these duties, although not easily performed, are self-
explanatory.

15. The Corps requires the contractor's QC controls to
include at least three phases of inspection for all
definitive features of work. The first phase is a
preparatory inspection which is performed prior to
beginning work on any definable feature of work. It
includes a review of contract requirements; a check to
assure that provisions have been made to provide required
control testing; an examination of the work area to
ascertain that all preliminary work has been completed; and
a physical examination of materials, equipment and sample
work to assure that they conform to approved shop drawings
or submitted data and that all materials and/or equipment
are on hand. Subsequent to the preparatory inspection and
prior to the commencement of work, the contractor is to
instruct each applicable worker as to the acceptable level
of workmanship required in his CQC plan in order to meet
contract specifications.

16. The initial inspection, or second phase, is performed
as soon as a representative portion of the particular
feature of work has been accomplished. It includes
examination of the quality of workmanship and a review of

control testing for compliance with contract requirements, use of defective or damaged materials, omissions, and dimensional requirements.

17. The follow-up inspection, or third phase, is to be performed daily to assure continuing compliance with contract requirements, including control testing, until completion of the particular feature of work.

18. The aforementioned activities cover the process through the construction phase but, the Corps QC/QA system does not end with the contractor's project completion. A joint inspection between the U.S. Army Corps of Engineers and the user is performed prior to user acceptance. Following correction of any deficiencies found during the joint inspection, the project is accepted and turned over to the user. Four and nine month warranty inspections are performed on the project after the user has taken occupancy. If deficiencies surface after occupancy, the Corps works with the user to correct the deficiency. These fixes may be in the form of the user fixing the defect, the Corps assisting in attaining money to fix the problem, or the Corps enforcing the warranty. The Corps tries to make the project turnover a handshake situation and not a handwash.

19. The Corps rates a contractor's performance upon completion of a project. This includes rating the architect-engineer, if applicable, as well as the construction contractor. The construction contractor rating areas are: (1) quality, (2) safety, (3) labor, (4) time, (5) management, (6) overall. These ratings are forwarded to the procurement office where they can be referenced for future contractor responsibility determination for contract awards. These ratings form an important knowledge base. They can serve as a basis to provide the Government with qualified contractors who produce quality projects. They can also help the Government avoid awarding contracts to unsatisfactory contractors.

20. To close the construction loop and avoid repeating the same mistakes, the Corps has a design/construction feedback system. The system involves collecting information from inspections, change orders, value engineering submittals, maintenance problems and review comments. This information is then made accessible to the interested parties. The system is designed to consolidate and organize the data collected, diagnose/analyze the data, prescribe fixes and then publish the information in a single source. Although the system does not operate as optimumly as hoped, it is a start to closing the information loop. The system offers access to many people's varied experiences and construction site learned knowledge. Through the use of the system, the Corps should be able to produce a better design, mitigate recurring deficiencies, improve construction quality, reduce cost growth and provide better user satisfaction.

21. Unlike private industry, the U.S. Army Corps of Engineers cannot arbitrarily choose the contractor it wishes to have. Rarely can the government choose a firm which designs and constructs the entire project because it likes the firm. Also, to avoid any possible conflict of interest, the designing AE for a Corps facility is not permitted to construct the facility (There are a few exceptions). Therefore, to ensure the government is getting its money's worth, special systems such as those listed above in the QA/CQC system have been put in place.

22. Depending upon the work load of the U.S. Army Corps of Engineers and the complexity of the project, the Corps may design the project themselves or hire an architect-engineer firm (AE) to do the design work. Unless the complexity of the job necessitates the AE's presence on the job-site, or it is to the government's economic advantage to retain the AE, the majority of the AE's job is complete before construction begins. The Corps answers most design questions during construction. If questions or problems in interpreting the design arise during construction that the Corps cannot adequately answer, the AE is queried. If the problem is due to design error or omission the AE is responsible for correcting the problem at no cost to the government. The AE is held accountable for any additional costs to the Government arising from its failure to provide a quality design.

23. When an AE is needed to design a project, the AE is selected by a selection board from a list of qualified AE's based on his qualifications applicable to the project to be designed. A price for his services is then negotiated. If a reasonable price cannot be negotiated, another AE is selected and the negotiation procedure starts over. This differs from the process of obtaining a construction contractor. As described before, the construction contractor is chosen based on the lowest bid submitted to the government contingent on the contractor being found to be a responsible contractor.

24. The construction contractor is responsible for all activities necessary to manage, control and document work so as to ensure compliance with the contract plans and specifications. Although, the contractor does have the incentive on many U.S. Army Corps of Engineer contracts to perform value engineering. If the contractor can find a more economical, better or improved material, design, etc. than the one specified in the contract, he can submit the one he found for Corps approval. If approved, the contractor is rewarded for his value engineering effort.

25. The QA/CQC system developed by the U.S. Army Corps of Engineers is equally effective if adopted by smaller

companies. The only drawback is that many of the
construction contractors performing Corps work add 10% to
the bid price to cover the cost of the upfront QC. In the
past, contractors have failed to recognize the potential
savings a properly functioning QC can make.

26. Some of the larger construction contractors in the
U.S. have their own internal QC system. But, most
contractors do not provide the services the U.S. Army Corps
of Engineers QA/CQC system requires unless it is stated in
the contract and therefore added to the bid price. The
realization of the need and benefits of a formalized QC
system are beginning to be recognized across the U.S. In
quantitative terms some have estimated the cost of poor
quality to be at least 7.5 percent of total project costs.
Until the cost and benefits are established, many
contractors will remain reluctant to institute a good QC/QA
system because of the obvious upfront cost of a QA/CQC
program.

27. The Construction Industry Institute (CII), as well as
the American Society of Civil Engineers, (ASCE), has taken
steps to promote the construction of quality projects. CII
is looking at various aspects of QC including defining
design and construction quality costs. ASCE has produced a
manual of professional practice, "Quality in the Constructed
Project, A Guideline for Owners, Designers and
Constructors". In the ASCE manual the importance of
concepts and practices to improve quality in constructed
projects is discussed along with roles and responsibilities
of all the participants from the owner through the
constructor.

28. The economic situation has forced the construction
industry to take a serious look at costs and tighten their
belts. The result is closer to optimum designs and optimum
productive labor forces. Designs are no longer overdesigned
enough to allow large variations during construction and
still produce a "quality" project. Since designs are
tighter, the conformance to these designs must be tighter.
The importance of quality in the construction process has
thus been greatly amplified. This realization is spreading
across the United States and with it is coming a more
organized approach to QC/QA implementation. With the effort
of CII, ASCE and others like them, QC will be seen in the
near future more as a commonplace necessity rather than a
paid for luxury. The result should be better designs,
mitigation of recurring deficiencies, improved quality,
reduced cost growth and improved user/owner satisfaction.

4.   Quality assurance — it will improve things won't it?

D. H. KETTLEWELL, MIStructE, National House Building Council

QUALITY ASSURANCE - IT WILL IMPROVE THINGS WON'T IT?

SYNOPSIS. NHBC sets standards for house builders to comply with. The process of house-building is too complex to gain total conformity with specification and consequently NHBC is appraising quality management to gain improvements. This paper asks, whether improvements can be made and provides answers from NHBC experience to date.

EXPECTATIONS.

1.   This paper expresses a view from NHBC experiences. It is a customers' view. Quality to the customer means better assurance, consequently quality assurance means guaranteed to be better. Whether the introduction of quality management and quality assurance will in reality mean a product or service is guaranteed to be better depends whether the customer's expecations have been established and met.

2.   A customer may expect higher standards in the product and if this is not achieved the customer's wishes will be unsatisfied. There is nothing to prevent the adoption of higher standards through Quality Management but this is not normally the first objective.

3.   The customer may be satisfied with the standard of the product but require greater consistency. As greater consistency is one of the normal objectives this particular aspect is likely to be achieved. The customer may be interested in greater efficiency leading to better services and reduced costs. Greater efficiency should be obtained but a better service will depend entirely on whether efforts are made to identify deficiencies and remove them. Reduced costs to the customer may not be realised if savings are absorbed by unnecessary bureaucracy for greater profits.

4.   One of the greatest concerns of customers is that they should not receive a defective product or service which gives rise to costly problems. Errors however are most

difficult to eliminate because as examination will show they have numerous sources, often outside the company concerned, and consequently a great deal of work and effort is necessary to identify prime sources and take appropriate action.

NHBC EXPERIENCE

Role of NHBC

5.   NHBC is the voluntary self-regulatory body for the house-building industry.

6.   The NHBC comprises nominees of all relevant interests in housebuilding but is not dominated by any one of them. It has no vested interest except in fair play and has no objectives save the promotion of quality and good standards in house-building and the protection of home buyers.

7.   NHBC is in the business of improving the quality of house-building and in identifying defects and reducing claims. Although broadly 1 in 100 dwellings give rise to claims for major structural defects on NHBC's insurance, these claims cost between £8.5 - £10 million a year. About 1.5 million homes are covered by NHBC warranty. But we are also aware that these claims are only a small part of the real cost of waste, inefficiency and a lack of quality. We firmly believe that a cost effective approach to the management of quality will have major benefits for house-building.

What causes defects and why is NHBC unable to totally eliminate them?

8.   It is a mistake to say that all defects are caused by lack of skill, attention to detail or supervision on site.

9.   Based on NHBC inspections and meetings with management and staff of house-builders, NHBC have identified examples of management's purchase of cheap, defective or incorrect materials; materials incompatible with design requiring alterations or replacement on site; site supervisors working on out of date drawings or no drawings at all and not being notified by designers of amendments to drawings. For these reasons we believe a cost effective approach to the management of quality will have major benefits for house-builders and house purchasers.

The need to persuade builders to implement and adopt quality management techniques

10.   We have to be mindful of the different needs and expectations of house-builders, to recognise their different sizes and priorities. We have approximately 26,000 builders on NHBC Register; from the largest, building thousands of houses a year, to the smallest which might only build one, say, every other year. There are nearly 9,000 builders who build only between 1 and 10 houses a year. On the other

hand, the top 50 builders build broadly half of all houses
built in this country. Collectively, the industry is
currently building about 200,000 houses a year. To gain
builders' support and to provide guidance to persuade
builders to adopt management techniques, which they may see
as bureaucratic, is a mammoth task.

NHBC decided to adopt Quality Management techniques for itself

11. NHBC started with an experiment with builders
reported later in this paper, but quickly took the view that
if we were to be successful in encouraging house-builders to
adopt quality management, we had to learn and hopefully
benefit from adopting these techniques ourselves. We are
now well advanced in introducing a quality management
programme within our own organisation.

12. We started in February 1986 with a weekend seminar
for senior staff. This was the first major event marking
the introduction of quality management within NHBC. The
Project Mangement Team was established to manage and
coordinate all quality management activites and syndicate
groups, defined corporate policies and objectives. During
the spring and summer all regions and departments took part
in three exercises designed to measure NHBC's performance.
First, shortfalls in the services we provide to our
customers were identified. Second, problems caused by
others, either colleagues withing the organisation or
external customers such as builders or home buyers, were
listed. Third, time spend firefighting, in other words time
spent which could have been avoided, was recorded. Amongst
other things, the results revealed that NHBC was not
sufficiently helpful in explaining its scheme clearly.
Builders often filled in forms wrongly which resulted in
NHBC making time consuming corrections. Furthermore, it was
estimated that an average of 1.5 to 2.5 hours each day was
spent by each employee firefighting at head office.
Communication, or lack of it, was the main problem. Forms
were either not being completed or were wrongly completed
and considerable time was spent in dealing with queries.
The NHBC scheme has now been simplified and better
explanatory notes provided. A new improved version has been
issued. Our forms have been re-designed and better guidance
notes produced.

13. In October 1986 we held a core group seminar and
representatives from 4 regions attended and were involved in
syndicate exercises to identify inspectors' and regional
managers' activities and problems.

14. The first of our Quality Management Newsletters was
issued to staff in December 1986 explaining to everyone why
quality management has been adopted and gave information of
progress.

15. In January 1987 at our Annual Field Staff Conference, collective expertise of the whole field staff was utilised to provide a valuable contribution towards the formulation of a new technical standards handbook. Syndicated groups were each involved in brainstorming separate handbook subjects.

16. A month later, our problem solving programme commenced under which all regions and departments received training in problem solving techniques, which resulted in some 40 proposals being formulated to improve actual working practices.

17. In May 1987 a full time appointment of Quality Management Coordinator was created to alleviate the increasing coordination requirements.

18. The first of our plan presentations took place in June 1987, when 8 proposals arising from the problem solving training programme were presented by Regions and Departments to the Project Management Team to be considered for implementation. One of these plans, an improved inspection card, was introduced in January 1988. This was followed by further issues of our NHBC magazine on quality management again to keep staff informed of progress and to provide an opportunity for them to express their views.

19. In July 1987 NHBC's objectives were arranged into 7 functional groups by the Project Management Team, intentionally crossing traditional regional and departmental boundaries. A chairman and a small interdisciplinary team were appointed for each group. They are responsible for managing projects and formulating proposals for improvements. We called them Group Action Teams. They take into account the results of any relevant existing work carried out in regions or in departments. All proposals are submitted to the Project Management Team for decision.

These 7 objectives for customer satisfaction are:

Management of quality
Marketing and publicity
Standards, prevention of defects and claims
Protection of the home buyer
Functions, organisations and procedures
Staff, including training
External suppliers
Difficulties were experienced as the work cut across responsibilities of office managers. Considerable revision of the work of the activities of the Group Action Teams has had to be undertaken.

20. The second of the plan presentations took place in September 1987 when a further 9 proposals were presented to the Project Team. As a result a new potentially more effective approach to the way inspections are carried out is now being considered. A policy statement including the

purpose of NHBC was published in our Quality Management magazine and our objectives were made available in all main offices for staff reference. All approved proposals will be implemented gradually in a sensible order to achieve cumulative improvements. An outline network diagram of the quality improvement programme has been agreed by the project Management Team as a working document. Some networks are now being added as more projects get underway, so that an overview of all work is provided to assist in identifying priorities and coordinating projects.

21. To date we have improved internal technical query procedure, transfered our registered office from London to Amersham, improved the speed of disciplinary procedures, reviewed the brief for our Information Department, improved builder application form with explanatory notes. We launched the "Buildmark" in 1988 being NHBC's improved system of customer protection. All the legal documentation was written in plain English. A Corporate policy has been adopted to establish a style for all NHBC publications.

22. The revision of the technical handbook has commenced its final drafting stage following intensive background research involving the field staff. We took the view that the handbook was a field staff tool and that they needed to be involved in every aspect of its preparation. Together we have identified the prime cause of the major incidence of defects, the problems of greatest concern found on site and in designs by inspectors, engineers and plans examiners, the problems of concern in particular regions and concerns for the future. We created a list of building functions from the purchase of land to handover of the completed dwelling as a checklist to help ensure all important functions were covered and as a means of storing information in a logical order. We have identified over 40 different sections for attention. Each section will be dealt with separately in a Practice Note. There will be chapters to deal with design, materials and workmanship and each chapter will contain performance standards and useful guidance on the ways to meet the standard.

23. We are also undertaking work to revise the procedure for dealing with late registrations of dwellings by builders. Major projects now being implemented include establishment of publications and stock control system, new inspection card system. A devolution pilot scheme was undertaken in one of our Regions in which we examined the implications of devolving responsibility for certain services, such as claims handling and registration, from our head office in Amersham to our regional offices. The aim being to make NHBC staff more available to the customer. This pilot scheme was successful and resulted in the project being extended to all NHBC regions. Complete devolution should be achieved during 1989.

24.   We have also undertaken a corporate data analysis, to provide management services information and technology planning.

25.   Further proposals being formulated or completed include new and improved methods of inspection, review of our administration procedures, production of induction training packages, review of communications, production of a building control explanatory pack for builders and staff.

26.   All staff from the bottom to the top are involved, improvements are now being implemented and seen by all to be effective. A new approach by staff to their job is apparent. They are questioning what they do, why they do it and how it can be improved. They are better providing their customer, be they colleague or outside the organisation, with the service they want at the right time and getting it right first time. We have learnt and are learning about quality management. We are benefiting from the improvements it is providing and armed with this experience and knowledge we feel better able to encourage house-builders to adopt and implement similar techniques.

27.   We have done all this not to receive a certificate to show we are competent or to gain national accreditation as an approved organisation to certify others, although we may eventually seek this, but because we want to know all about the management quality - warts and all - so that we can improve NHBC and from our knowledge help the whole of the industry to achieve better quality.

28.   It will be apparent that what NHBC has undertaken is a total restructuring. It was, and still is, a massive job. With hindsight it could be said that it would have been easier to restructure first and set up quality systems later. Although not specifically planned this way that is exactly what has happened. The impetus created by early educational exercises gave rise to the identification of deficiencies which in turn gave rise to restructuring and numerous improvement projects. We still have not laid down a quality system. We get the feeling that this learning curve will take us to ... well the sky's the limit so why not expand to the stars? 1992 opens the first part of the route.

NHBC initiative with house-builders

29.   We believe that before we can introduce quality management to the industry as a whole, we must first of all learn and establish what the best and most cost effective ways of achieving quality objectives are across a very wide spectrum of size and type of builder. We of course agree with the wisdom and value of the principles of BS 5750 but nevertheless have to bear in mind that for many house-builders the very mention of this document tends to alarm them about the possible increase in bureaucracy and paperwork, which they believe would fetter their freedom of decision and action and add to their cost. We consider that

the first job is to break down such misunderstanding and
apprehension by introducing quality management in a way
which will show that it is no more than good business
practice and commonsense.

30.   We have also to take into account the nature of the
house-building industry, the many and varied people
involved, the many inter-related products and services
involved and the fact that much of the major work is subject
to the vagaries of the weather.

31.   We first of all embarked and funded a pilot scheme to
ascertain whether the quality management approach could be
tailored to the requirements of the house-building industry.
The very positive approach from the builders showed that it
could.   First of all, in discussion with our consultants and
the participating builders, the following objectives were
agreed:-

   to design and implement a quality management system at no
   extra cost to the company and aim at net savings, in
   particular to:-

   (a) reduce the incidence of defective work at all stages
       from market research through to after sales service;
   (b) reduce the incidence of customer complaints;
   (c) increase customer satisfaction;
   (d) enhance company image;
   (e) increase the morale of employees and their commitment
       to the success and profitability of the company.

32.   The NHBC invited about 100 companies to a meeting in
London to discuss the proposed pilot scheme and to see if
there was sufficient interest for the project to be
launched, with the initial intention of 5 companies
participating.   A 100 companies were selected from those
building between 150 and 1500 houses a year.   Eighty
companies accepted our invitation and, following the
presentation in which the nature of the quality management
scheme and the level of commitment required from each
participating company were described, thirty seven companies
wanted to participate.   Because of this high level of
demand, we agreed to fund 10 companies instead of the
original 5.   The 10 were selected to give a wide
geographical spread as well as a wide range of organisation
type and size.

33.   The foundation of the project was that each company
reviewed all of its management procedures that affect the
quality of the final product or affect the cost or ease with
which quality was achieved.   Each company established a
Project Management Team which was representative of all
parts of the company and over the 12 month period of the
project received 15 visits from the consultant.   Our own
field staff attended most meetings between the consultant
and the companies.   Not surprisingly, most companies were
sceptical at first but the fact that they were participating

showed that they were open minded, an essential pre-requisite of success for the scheme. During the course of the scheme two companies were involved in takeovers, one company was involved in a management buyout and three others experienced extensive company reorganisation. Thus during the 12 months the scheme ran, 60% of the participating companies experienced considerable disruptive managerial reorganisation in one form or another. However, despite these upsets, only one of the 10 companies dropped out of the scheme.

34. At the end of the 12 month programme, there was unanimous agreement from the remaining 9 companies that the scheme had been a success. Without exception, all companies were able to outline the benefits they had received and indicate ways in which they proposed to restructure their administrative procedures and improve liaison. An important feature of the project was that it turned traditional thinking on its head. As pointed out earlier, traditional thinking has it that because problems occur on site, they are actually caused on site. Builders assume that the problems are caused by factors beyond their control, such as the weather, the multiplicity and location of sites and the high turnover of labour. But, as our research has indicated, the basic premise of this pilot scheme was that many problems that occur on site are caused earlier in the development. For this reason the project in each company did not concern itself in what happened on site until head office operations had been thoroughly reviewed and this took about 6 to 8 months. By this time, benefits were beginning to appear on site, although not as quickly as we expected and would have liked. More about this later.

35. Some of the companies involved have already started to involve their suppliers and sub-contractors in quality management with beneficial results to both parties. This is a trend which must be encouraged. The need is for many more companies to adopt the quality management approach and to involve their suppliers and sub-contractors in the programme. One of the outstanding benefits of the pilot study was the way in which it built on the willingness, the desire for people in all parts of the company to work together if they are shown by top management that they are willing to take the lead by demonstrating their total commitment to doing their jobs right first time. This pilot scheme clearly demonstrated that the house-building industry could greatly benefit from the adoption of quality management techniques. The scheme also demonstrated that the industry is as managerially advanced as any other industry.

36. Based on the experience we gained on the first pilot study, plus our own in-house experience, we have embarked on a second pilot scheme involving 15 companies of different sizes and types. Based on one of the lessons we learnt in the first pilot scheme, we intend to improve the monitoring

of the before and after situation on site and having identified problems on site, trace them back to their origins within the company's head office. This is an aspect that we did not deal with adequately in the first pilot study.

37. Based on our experience to date, plus discussions with major house-builders, it will be inappropriate to encourage house-builders to adopt and implement quality management by slavish adherence to all the recommendations of BS 5750 and develop total quality management systems. BS 5750 concentrates on ensuring that the builder complies with his specification, not that the specification is "right". We take the view that if the most critical aspects, as identified by customer's complaints and costly errors or major change, can be identified and improved it will be a major step forward and a springboard for further improvement. However, NHBC readily appreciates that it was not intended that BS 5750 should be applied or used in such a manner but the principles it contains can be followed. We are aware that the selection of appropriate elements of the Standard and to what extent they are enforced depends upon the nature of the company and/or product or service. From NHBC's experience, we believe that it is first necessary to agree that most builders would benefit from simple practical guidance and that even the best would benefit from critical self-appraisal.

38. We believe that the house-building industry should itself produce a standard manual for quality. This would explain and expand in simple form those elements and principles of BS 5750 which are relevant, plus any other guidance which we think would be of use. With the help of house-builders, NHBC is in the process of producing such a document for our own industry, so as to help builders to help themselves and introduce a more structured approach to those aspects of management and organisation which effect quality. As a result of this approach, we could then ask builders to implement the manual for quality on a voluntary basis.

Well, has anything been proved?

39. Undoubtedly every employee of every company and every customer can benefit but NHBC experience is that it is tough going. The educational process is highly beneficial enabling all staff to solve everyday problems in a systematic and logical way. It is, however, important to take account of the power that can be released from staff initiatives and to some extent it is necessary to temper the enthusiasm developed so that the overall effort can be controlled to the common good. It is to be appreciated that in a large organisation even a small departmental initiative can have far reaching consequences. It will be noted from this paper that NHBC had to establish a corporate function plan to help understand the implications. Because of this

it is probably helpful to tackle isolated aspects first so as to avoid highly complicated knock-on effects and gain early successes. This is particularly important if the suppliers and customers are to be impressed. Unfortunately, NHBC inspectors did not perceive any improvement in the site work of the builders who took part in the first experiment but this was perhaps because the site problems were not dealt with adequately or sufficiently early to become apparent. As indicated in the paper the second experiment is dealing with this. Once the quality systems are in place it becomes very much simpler to tackle problems but in the early stages problem solving takes time and the lack of progress can be frustrating.

40. The benefits internally became apparent more quickly and are particularly noticed in improved efficiency and enthusiasm. We have come a long way but it has also taken a long time and much still remains to be done. We remain convinced there can be dramatic improvements but to achieve them requires commitment, very careful planning, hard work and patience.

If you have any doubts try an experiment

41. Call together representatives of all the departments of your organisation, suppliers, anyone to whom you give service and your customers. Ask the questions:

What went wrong with your last project?

What caused the problems?

What could have been done to prevent them which could be applied to future projects?

Put aside a day, then a year and then a lifetime of commitment to continued improvement.

Discussion on Papers 1-4

MR KETTLEWELL (Paper 4)
The National House Building Corporation is currently in the process of adopting quality management procedures and as a consequence is undergoing a complete review of all its operating procedures and is introducing improvements. It is not easy to find the time to make radical changes but the NHBC is doing so because the advantages can be seen. The corporation knows that a major increase in efficiency can be gained, but more importantly is convinced change is necessary if the right service is to be provided to its customers.

At the same time experiments with house builders are being undertaken by which they are shown, with the aid of consultants, how to adopt quality management procedures. The results so far have shown increased efficiency and in this respect have been most satisfactory. Not one builder has said that the effort was not worthwhile. However, there have not always been improvements on site. The latest experiment is aimed at trying to produce data by which improvement can be quantified. The apparent lack of site improvement may be because the procedures did not reach site work operatives, or the improvements went unnoticed or because certain aspects for improvement were not targeted. It may just take longer for general improvement to show on site. Site improvements are possible because they have been seen through NHBC's Pride in the Job campaign which has gained general improvements and, in certain cases, dramatic improvements.

MR G. B. M. OLIVER, Bryan Oliver Associates
With reference to cost figures in the budget summary table in Paper 1, over what period were the Author's £27 150 of initial costs incurred? Also what is the approximate annual turnover of Design Group Partnership? Annual costs could then be related to turnover in percentage terms.

MR L. A. NACKASHA, Queen's University of Belfast
Papers 1-4 address the two major considerations for quality: the management of quality and the quality of management. Mr

Duncan broached the issue of quality cost. Large organizations, although they have related figures regarding the cost of their quality systems, have intentionally surrounded this issue with a cover of protection. Indeed the whole issue of quality systems is treated with secrecy, maybe for fear of copying or duplication of a company's QA system. These fears can be negative and detrimental to the industry as a whole.

QA systems for organization are, in real terms, as copy-proof as fingerprints. Although they may be copies and adopted in other organizations, they cannot be expected to produce the same performance. It is also wrong to think that the QA system can be bought and plugged in to the organization, like an electrical appliance, in order to work wonders. Quality is a belief, and to copy that belief strengthens it further. This is why the Japanese, when asked, are not afraid of discussing their QA systems. It is important to open boundaries and exchange expertise and opinions between construction organizations. It may even be a better practice to audit each other's system.

Papers 2 and 3 described the main ingredients of a good quality system. In general, the first three papers represented one side of the argument regarding quality systems, that is the management of quality.

Paper 4, however, presented the other side of the argument which is the quality of management. The Author demonstrated the emphasis of his organization on the end user of the final product. Indeed quality is about satisfying the needs and expectations of the end user. It is not about maximizing profit of promoters, because that is a major inducer of negative growth in the construction sector, although it may seem at first sight that the opposite is true.

Maximizing profit increases the probability of skimming over quality and safety procedures, thus increasing the chances of such mistakes being committed, which may lead to failures. Failures, in turn, are a major cause of legal disputes and insurance claims. These costs are totally unnecessary to the construction industry. They are responsible for a negative growth in the construction sector and a positive growth to other sectors such as the legal profession and insurance.

MR V. J. W. HOAD, Sir William Hall & Partners
Mr Duncan's Paper does a great deal to reduce the mystique which has tended to exist around quality assurance. It is most important that in a design office everyone from the Chief Executive down to the industrial staff should be enthused about the requirements for quality assurance.

In Halcrow the initial outlay over two years has amounted to about £100 000 but included in this were the review and revisions to the company's existing manuals for: the design office, contracts, project manager, resident engineer and

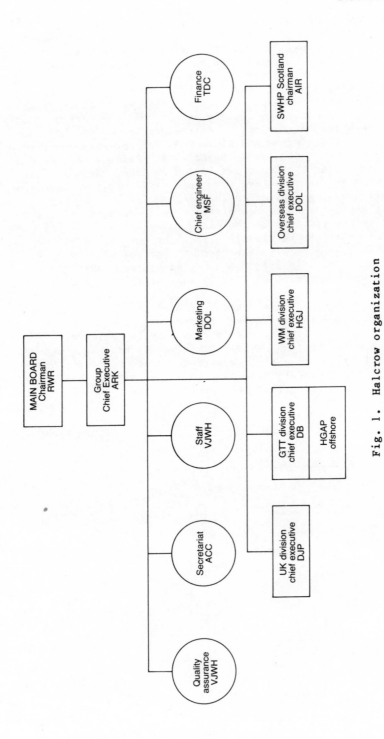

Fig. 1. Halcrow organization

architect, and safety, together with one for quality
assurance and procedures and a new visual standards manual.
The safety manual is currently being reinforced with regard
to site safety matters. The annual maintenance training and
system operation appears to be around £50 000 per year and
this includes a full time quality assurance manager.

Halcrow has taken rather longer with the development and
implementation of the quality system than the Design Group
Partnership, but the costs referred to appear to show some
benefit of scale numbers of 1000 in Halcrow when compared
with the 200 quoted by Mr Duncan. The Halcrow declaration
of policy for the Group was issued in 1986. It will take
Halcrow at least another year before all projects have the
quality system applied to them.

It may be wrong to denegrate too much of past efforts in
the industry. Some of the examples cited by Professor
McCaffer and Mr Kettlewell were about the 'state of the art'
and in others really about quality control which in
Halcrow's view is 95% of quality assurance. Having said
this the management of quality assurance is an absolute
vital element.

The company's organization is shown in Fig. 1. It
demonstrates the management's commitment to quality
assurance and similar charts exist for the Halcrow
divisions. There is a full time quality assurance manager
reporting to the quality assurance director and in the
divisions there are also some full time quality assurance
managers. Fig. 2 shows the quality policy, quality

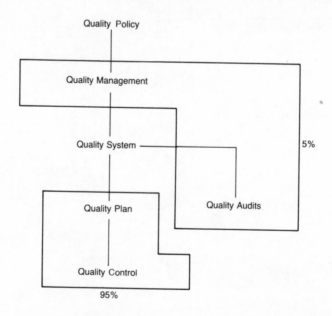

Fig. 2. Halcrow Group quality assurance

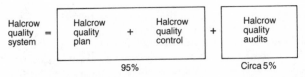

Fig. 3. Halcrow Group quality system: basic principle at project level

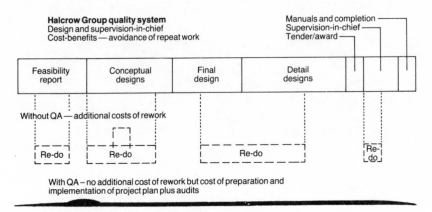

Fig. 4. Typical project quality system

management, quality system including quality audits, quality plan and quality control. The quality plan and quality control have in essence been in place, as with others, and form 95% of quality assurance input. The quality management and quality audits make up the 5% additional input.

The basic principle is shown in Fig. 3. in the equation at project level the quality plan and quality control again equate to 95% of the input with up to 5% for quality audits and quality management. Fig. 4 outlines a typical project where the application of quality management may obviate repeat and wasteful inputs which could amount to, say, 25% of the total work input with a saving amounting to as much as 20% in some cases. This is intended to demonstrate where benefits may lie.

There is no substitute for the integrity, competence, enthusiasm and conscientiousness of the people involved.

MS Y. ZALZALA, Technical journalist
Is QA offered automatically to clients, or is the onus on the client to ask for it? Is QA extended to the company's work overseas? How long are the QA project documents kept after completion of projects?

J. M. DUNCAN, Paper 1
In reply to Mr Oliver, £27 150 would be incurred over a minimum period of 12 months, more likely 18 months to two

years.   Approximate annual turnover is £3.5 million.
Ms Zalzala asked how is QA offered.   QA is a total
management system and is operated on all Design Group
Partnership's projects, irrespective of client preference.
It is our internal management system.   QA is used on all
design Group Partnership's projects at home and overseas,
for the reasons stated.   Project documents for QA are kept
for a minimum of 10 years, but more recently a 15 year
period is being considered.

5. Insurance, quality assurance and 1992

B. O. H. GRIFFITHS, TD, MA, ACII, Griffiths & Armour Insurance Brokers

UK Contractors' insurance portfolio examined: Insurance
requirements of Contract Conditions: "Design & Construct" - a
contrast of duties: Insurance for the "Design & Construct"
Contractor: UK Consultant's insurance portfolio examined: Why
not Structural Guarantees? UK & European Insurance portfolios:
Decennial Liability & Insurance: The effect of BS5750/ISO9000:
How will 1992 affect insurance?

Present Insurance Portfolio - UK Contractor

In addition to holding such insurances for his own business as
he feels it appropriate to hold the UK contractor will also
hold those insurances necessary to secure compliance with
contractual obligations. For example, to satisfy the JCT80
obligations for 'pre-1986-amendments' contracts it would
normally suffice to have in force EL and PL policies suitably
endorsed to provide the requisite protection for Principals,
and if necessary endorsed to comply with his own preferences
or his contractual obligations in respect of, for example,
property being worked upon. Additionally he will hold 21.2.1,
or 19(2)(a), cover, either on annual basis or one a 'one-off'
basis, for contracts involving a risk of damage to adjoining
property where such cover has been stipulated, and Fire &
Perils cover for buildings in course of erection, normally on
an annual basis. He will also doubtless hold suitable cover in
respect of his obligations for hired in plant and other
particular risks of the contract.

Where his contract is JCT80 incorporating the 1986 amendments
mere Fire & Perils insurance cover for buildings in course of
erection is not sufficient - hence the contractor will need to
hold Contractors' "All Risks" cover. Note the inverted commas,
for the policy is not without conditions and exclusions. Again,
in appropriate cases he will be required to effect 21.2.1. cover
in respect of his non-negligent liability for damage to
adjoining property

The civil engineering contractor working under the ICE Contract has no option but to carry Contractors' "All Risks" cover on account of the contractual obligation to effect such insurance.

When there is an event such as damage by fire, storm, flood or other insured peril the contractor will doubtless be instructed to put things to rights and will submit a contractual claim for the time and cost of the additional work which will be paid by Insurers of the appropriate policy: but when damage or loss occurs which appears possibly to result from a cause due to the design of the Works, perhaps a collapse or similar catastrophe, Insurers can be expected to invoke their policy's exclusion in respect of "a cause due to the Engineer's design of the Works"; events can then take one of several courses. For example, the contractor may be instructed to make good the works in which case he will put in a contractual claim; this may provoke the Employer into seeking to recover from the contractor on the basis that his erection plan or design of the temporary works was defective, and/or the Engineer with allegations of breach of contract and/or negligence. The overall design of the Works is, of course, the responsibility of the Engineer, but the erection plan and the design of the temporary works are among the Contractor's responsibilities.

"Design & Construct"

So far it has been assumed that the Works have been designed by the Client's Consulting Engineer and are being constructed by the Contractor whose tender has been accepted. There is, of course, also the Design & Build option, or the Design, Build and Finance 'turnkey' operation. Contractors who offer these services may undertake the design element 'in house' or they may call upon the services of an independent firm on the basis of 'design only'.

'Design & Build' Contractors not infrequently use the services of a professional designer and he is wise if he ensures that his conditions of engagement are drawn up with special care in such cases. The professional designer's normal legal duty is to exercise reasonable care, and it is on this understanding that his professional indemnity cover is written. If he was inadvertently to engage himself under an obligation of fitness for purpose he could find himself in difficulty with his Insurers just at the time he needed to call upon them.

'Design & Build' Contractors too are liable in law for their design: but, doubtless because they supply the materials and control the labour, they are additionally considered to be liable for fitness for purpose.

Insurance for the "Design & Construct" Contractor

It was once fairly common practice for such Contractors to hold Professional Indemnity policies, but in more recent years it has been realised that the classical Professional Indemnity policy is not really suitable for a Contractor. The unsuitability of classical professional indemnity insurance for the Contractor is exemplified in the matter of supervision – the insured contractor might wish to ascribe workmanship defects to his own negligent supervision in the hope of bringing the rectification costs within the ambit of his Professional Indemnity policy. It was for this reason that some Underwriters proposed and argued for the total exclusion of claims arising out of supervision. However such drastic action would have represented a degree of 'overkill' since the D&C contractor has supervisory functions, for example over subcontractors, for which he reasonably enough seeks and expects the protection of Professional Indemnity insurance.

There is also the whole question of the 'prior to handover' risk. In a conventional scenario if the contractor incurs expense due to a design error he will make a claim; if the designer won't issue a Variation Order the contractor will put in a claim for the extra work, and the law recognises his right to claim against the designer for breach of a duty of care owed to him. But what happens when the contractor is also the designer? He can't sue himself. So he seeks cover for his rectification costs "Prior to Handover". However the insurance market recognises clearly that in the conventional scenario a plaintiff must base his claim on alleged legal liability and run a risk in pursuing it. Insurers make it their business to secure and finance the best possible defence for their designer clients, and thus ensure that only the best claims are made to stick. Yet if a Design & Build contractor holds "Prior to handover" cover he merely has to show his Insurers that an insured event occurred in order to qualify for indemnity: in such a scenario Insurers lose the opportunity to put up a defence and find themselves on the opposite side of the table from their Insured.

Domestically the Insurers ask themselves whether they are being fair to their 'pure' designer clients, and indeed fair to their 'design & construct clients', if they pool their premiums together and pay claims from that pool regardless of the fact that their legal duties, and the risks, differ.

The practical, legal and insurance issues call for specialist broker analysis and guidance in the placing of these insurances and the handling of incidents and claims which fall to be reported.

Present Insurance Portfolio – UK Consultant

As with the Contractor, so with the professional consultancy practice insofar as its portfolio of domestic insurances for its own practice is concerned.

But unlike the Contractor who is obliged by contract conditions to effect liability and material damage insurances to protect the interests of himself and of the Employer, it is in theory entirely a matter for the consultant whether he decides to effect professional indemnity insurance or not. In practice, of course, the vast majority of professional Architects and Engineers in private consultancy practice do effect professional indemnity insurance, in some cases because their clients so insist. Whatever may have been the motive for effecting the cover it is so written as to protect the consultant and to provide him with the funds to meet his legal liabilities – it is not a material damage insurance like a fire insurance policy, legal liability has to be established and quantified before a claim is payable.

We considered the caution a Consultant needs to exercise in ensuring suitable conditions of engagement when he provides the designs for a Contractor who is offering a 'Design & Construct' package: the Contractor will have engaged himself in giving an implied warranty of fitness for purpose: if the Consultant unwittingly undertakes a Fitness for Purpose warranty, and if the project fails, the Consultant's Professional Indemnity Insurers might not be liable to give him indemnity. If a consultant holds himself out as having a special skill then he may risk being found liable for the exercise of something more onerous than the reasonable skill care and diligence of a competent practitioner: Greaves v Baynham Meikle is a case to study in this connection.

The moral is that Consultants should pay close attention to their Conditions of Engagement and have them clearly on the record for each of their appointments.

Why not Structural Guarantees?

Since Dr Powell-Smith is not able to take part as a Conference speaker I have expanded somewhat on legal aspects both in this and other sections of my paper.

The conundra posed by 'Design & Construct' have already been examined in some detail and they highlight the different legal duties owed by contractors and consultants.

Consultants are sometimes asked to give guarantees, such as Structural Guarantees: the request of itself shows a misunderstanding of the place of the professional. The professional is not an institution with access to significant funding – he is rather the 'skilled hired hand with specialist knowledge' hired in for his services by the client. If that client is a developer or an arm of the state he himself has access to sources of finance and it is appropriate for him to arrange such guarantees as he may want. If the client is a developer he seeks the rewards of speculation for himself and his shareholders - the professional is engaged for his skills and paid a fee, not a percentage of profits.

Too often Engineers' clients seem to have the impression that ever more onerous burdens can be placed on the shoulders of the Engineer thinking that his Professional Indemnity insurers can pick them up should the occasion arise: the Engineer can reasonably be expected to perform his own role competently and to pay the price of failure through his own PI insurance, but he is not structured or financed to handle Structural Guarantees, Collateral Warranties etc.   These are more appropriately sought on the financial markets by developers or from the public at large through taxation.   An Engineer complying with a request for a Guarantee would be treading a dangerous path.

The Consultant's duty is to bring to bear reasonable skill care diligence and competence, and that duty doesn't extend as far as giving guarantees.   The Consultant's professional indemnity policy is not intended as a security for guarantees; it is intended to protect the Consultant against awards of damages and costs, it is subject to conditions and exceptions and its renewal cannot be guaranteed.   And projects can fail for reasons outside the Consultant's field of influence or control: for example there could be damage by an insured peril, or due to the contractor's negligence, or due to the client's false economies - for example in declining a full soil survey of ground which proves to be unstable; or the owner or a subsequent purchaser may adapt or misuse the Works.

Comparison of UK Insurance Portfolio with European Insurance Portfolio

In countries which came under Napoleonic influence the fundamental problems are, of course, the same as those we face in the Common Law countries: however, if an analogy is helpful, they may be said to tackle the same mountain but from the other side.

In UK we work on the basis of first-and-second-party insurance covers where the cause of the loss or damage lies outside the control of the parties - fire and storm damage being typical examples.   Where the parties have control and it is possible to attribute legal 'fault' we work on the basis of attributing liability.   Insurance is effected against legal liabilities by the potential defendants.   Save in the case of liability for death or injury caused by defective products we do not make general use of the notion of 'strict liability'; in other words our system requires the plaintiff to prove that the defendant is liable and to prove his loss.

Decennial Liability & Insurance

In France, commercially the most important of the 'Napoleonic' countries, the builder and the professionals are strictly liable in law for minor defects (these include such matters as painting) for two years, and for major defects for ten years. This, the so-called Decennial Liability, is imposed by para 1792 of the Code Civil. In law and in practice defects manifesting themselves within the appropriate period are held to be the responsibility of the builder and the professional jointly and severally unless they can prove otherwise. The presumption in law is that they are liable - if they wish to contest it the onus of proof is on them.

There is a structure of interlocking insurance policies which provide for the Decennial legislation. The consultant has his cover for his decennial liability, the contractor has cover for his decennial liability, and the building owner has 'material damage' cover for problems or defects of a decennial nature. Should a loss occur the owner will notify his Decennial Material Damage Insurers and they may have subrogation rights against the (known) Decennial Liability Insurers of the designers and contractors.

The existence of Decennial Liability in France does not mean that there is no need for Consultants to effect and maintain Professional Indemnity cover. Decennial insurance covers only physical damage and defect in the building: it does not address such problems as claims from Third Parties who may suffer death or injury or financial loss, nor does it address the liability of the Contractor and/or the Consultant for extraneous matters which may form the subject of a claim by the building owner, such as inadequacy of a heating system or delay in the production of drawings. Hence the Consultant in France will carry not only insurance for his Decennial Liability, but also conventional insurance for Professional Negligence.

Decennial insurance represents another burden of cost, but it does give the owner his own 'material damage' policy upon which to claim and thus obviates the need to prove the liability of another party before being recompensed. It is also noteworthy that for major projects the available Professional Indemnity limits often represent but a small fraction of the cost of the largest possible loss.

Quality Assurance and Insurance - BS5750/ISO9000

Clearly any measures which serve to improve a risk are welcomed as much by the Insurers as by those who adopt them. If wide adoption of the appropriate part of BS5750 becomes the norm within Consulting firms and in the contracting industries, and if as a result the cost and number of claims falls, then this will be reflected in the premiums which Insurers require.

The acid test will be whether the number and cost of claims drops. Insurance is a commercial business and Insurers are in business to make a profit. Indeed, if it were not reasonably profitable then its security would be in doubt and without security its value would be little.

It is possible that in those periods of the underwriting cycle which pass for a 'soft market' some Insurers may offer a small discount to policyholders who have an accredited Quality Assurance scheme in force. However these can only be token gestures, for the obligation upon Insurers is to collect sufficient premium (and accrued interest) that they can meet the cost of the claims which policyholders make.

Rate-setting in the insurance business may initially be based upon theory, but it is ultimately determined by claims experience. Any premium reductions on account of Quality Assurance will be those which have been earned by its successful application in terms of claims reduction. The mere existence of an accredited Third-Party-audited scheme will not of itself lead to automatic premium discounts.

There is a possible hazard which faces firms where a QA system is in place. It concerns the situation which might arise if a particular action taken within the firm should prove to have been demonstrably in defiance of the firm's own QA Manual: without QA in place it might have been possible to have defended the action, but with QA such a defence would be precluded. It will be interesting to see whether this comes to pass, and indeed what other unintended downsides QA may bring in its wake.

How will 1992 affect insurance?

The concept of Open Markets inevitably means that not only will UK Brokers and Insurers have free access Europe-wide to those risks which are not reserved for domestic Insurers, but also their Brokers and Insurers will be entitled (hopefully within similar limits) to have access to the UK market. The concept of reserving certain risks to domestic insurers may appear to be a somewhat restricted view of 1992 but it seems at the time of writing to be as probable a scenario as any.

In the countries of Continental Europe many UK Brokers and Insurers, either directly or through associations with locally-based operations, are established members of their respective market-places. Within the UK market the number of European-owned Insurance Companies addressing the retail market is small, but there are cross-shareholdings. Several of UK's major Brokers are now wholly or partly American-owned and vice versa, but links with European Brokers are less numerous.

The choice of policies and the manner of their arrangement in different countries has evolved from the need to meet the law and practice in those countries: hence a French-owned Insurance company offering Decennial cover within UK there has found that there is little interest on the UK domestic market. Given the problem posed by the many languages and legal systems in the Community which can hardly be 'harmonised' at a stroke it seems probable that, insofar at least as the great majority of insurance risks are concerned, the insurance markets will continue to operate substantially as before for several years to come. However the constraints which tend to favour the selection of a 'domestic' insurer for a 'domestic' risk will gradually be eroded and, starting with the larger risks, we can expect to see Continental-based insurance concerns competing with UK-based competitors for UK risks and vice versa.

Perhaps more interesting will be the attempts to harmonise conditions of contract and conditions of engagement, the legal duties of the parties, corporate and personal taxation, and the panoply of matters which have been within the domain of individual nations but which make for a less-than-level playing field across Europe. And perhaps by then there will be additional players on that field who have in theory let us play on their fields but in practice placed numerous ingenious handicaps in our way. Who knows, we may be persuaded to codify our common law - or even succeed in exporting the notion that an unwritten Constitution has its merits!

6. The liabilities of the professional designer

J. M. ANDERSON, DA, FRIAS, RIBA, MSIAD, Bickerdike Allen Partners

SYNOPSIS. In the creation of a building, the professional designer fulfils not one but several different roles, each with its own peculiar risks and responsibilities. Whether the process of Quality Assurance can help him reduce these risks and deal better with his responsibilities depends upon a recognition of precisely what these roles are and how they relate to those of other people involved in the enterprise. It may be that Q.A. will be more easily applied to some aspects of the designers work than to others. If we use too broad a brush the casualty may be good design itself.

1. The process of creating a building doesn't begin or end with a designer's drawing. It starts with a client and ends with a building contractor. But the designer has a role, or rather a series of roles throughout all the activities from concept to completion. Each role brings with it different responsibilities and liabilities. The question I want to address here is whether the techniques of Quality Assurance can reduce the risks not only for the designer but for everyone involved in the lengthy and complicated business of building.

2. Some of the designer's responsibilities are set out in the terms of his engagement and in the practice documents published by his professional body; others are in the building contract; some are not set out or defined anywhere.

3. His various activities overlap and fluctuate in the course of a project and he will rarely stop to consider whether what he is doing is set down explicitly in some document. But since Quality Assurance has to do largely with making the implicit more explicit, it is inevitable that designers will begin to see their roles somewhat differently as they become increasingly subject to QA procedures.

4. This change is not altogether unheralded. It has been

preceded for some years now by the publication of an increasing number of practice notes and handbooks and by the effects, direct and indirect, of building litigation which have forced us to review the ways we do things and sometimes to pay for having done them with less care than the Courts thought our clients were entitled to expect.

5. Whether this will make us more willing candidates for QA assessment is not at all clear but I will suggest that it would be regrettable if it results in designers becoming so dominated by their practice procedures that their more creative contribution to building is lost or diminished.

6. The designer's functions can be teased out into six parts which are sufficiently distinct to allow us to see how each may be influenced by the process of Quality Assurance. First there is the CONCEPTION of the building, closely followed by the COORDINATION of the design team.

7. In bringing forward concepts of space and form which meet the case aesthetically, functionally, constructionally and economically, the chief procedural instrument is the Brief. In times when patrons and builders had a common language and building technology was old, a Brief could be very brief indeed. The patron could expect that a simple demand for a church to the style of another - but larger or smaller - or a mansion grander than his neighbour's would bring forth a design that met all his expectations of firmness, commodity and delight. (If it didn't the result for the designer might be simple, swift and even fatal.)

8. Today's building brief starts large and grows. Before the design team and the client body can be satisfied that what is intended matches what is needed within all the constraints, the briefing document may be a very large volume.

9. Although achieving beauty in building is generally thought of as his foremost duty, I have not known of any case where an architect has been held to account for an aesthetic shortcoming in his design. It would make for very interesting litigation. On the other hand, I have known of cases where functional deficiencies (stairs seem favourite amongst things that don't always work) and designs that far exceeded the client's budget or time-table have become matters of serious contention.

10. Obviously in the brief-building exercise the designer is not alone but he may still be uniquely responsible for telling others, including his client, just what and when things have to be decided if the process of developing the brief, and its gradual transmogrification into a design, is not to go seriously off the rails.

11. Of course many of today's clients are very experienced in matters of building and may even field their own team of professionals to help the designer understand their precise and detailed needs, but this is unlikely to save him if things go wrong at the conceptual stage. He alone is supposed to bring it all together in the right way, at the right time and at the right price. Especially where heavy commercial considerations are at stake failure at this stage can be very costly.

12. The conceptual role runs into and over that of COORDINATING the design team. Much arrogance and resentment surrounds the architect's claim to leadership of the design team but it is an unnecessary squabble. Without a design a design team doesn't work. At every stage of developing the general arrangement or the detail the other members of the team turn expectantly to the professional designer. If he says nothing, nothing happens. It is only when he unrolls his sketch or drawing that the team's work can continue.

13. This initiating role places the professional designer in a position of special responsibility. The same token that makes him central in the team activity makes him sensibly the only person who can define and coordinate the contributions of the other team members. This is not the same as being responsible for what they do but it does entail checking that they are doing, in a careful and timely way, what they are specially equipped to do. Without this connective tissue the fabric of the team can unravel with dire results for which the designer, as leading consultant, will be held largely responsible.

14. So much for the form of decision making. What of its content, the TECHNOLOGY and PRODUCTION INFORMATION that flow from it? Nowadays when we think of building failures and litigation it is primarily in terms of leaky flat roofs, spalling walls, fractured floors, condensation or rising damp but surprisingly it is not necessary in today's fast-moving science and technology for an architect to be omniscient. It is only required that he should be careful. Today's buildings are vast conglomerates of new materials put together by new methods. Because of this and the new demands we place on our buildings none of today's ways of building is tried and tested because no test known to man simulates faithfully the effects of time, environment and the natural elements. If it were not for the incredible optimism of designers, materials manufacturers and builders no one would dare build anything these days, and if it were not for the largely forgiving nature of building technology, our technical failures would be far more numerous.

63

15.   Technologically the architect's duty is to be aware of the current state of the art, however shifty and incomplete it may be, and to ensure that everyone on his team does his best within the uncertainties of fast-moving times. In matters of technology especially, it is possible to be wrong without being found negligent; it is not, however, possible to be found negligent without being wrong.

16.   One cause of negligence is undoubtedly pressure of time and in the production of information that pressure can be increased greatly by duties imposed upon the designer by the building contract. Not long ago it was most unusual to find that time was "of the essence of the contract." Now it has become the norm and as a result the architect must fulfill many of his duties under the contract to the contractor's programme, however imprecise or arbitrary that programme may be. This has created a whole "claims industry" ready to pounce should he fail to manage on time the flow of information and decisions throughout the construction period, and during crowded and lengthy site meetings.

17.   Conflict of interest is built into his role as judge on the Contractor's claims for extensions of time, loss and expenses and any other matters where it can be said that the architect has influenced the progress of the works. One false step and he can find himself choosing between facing an arbitration under the Contract because he refused a claim or facing litigation by the client because he approved it.

18.   Usually the contract requires the designer to inspect the works and to satisfy himself that the contractor has fulfilled the practical terms of the contract before he receives payment. This inspecting of works and certifying of monies places a heavy responsibility upon the designer for which he is not always temperamentally suited so he has to rely on clerks of works and quantity surveyors. But these collaborators are finally only extensions of his role and the most difficult problems will still come back to him for decisions.

19.   For example, ordering opening-up works for inspection may result in a costly delay and I've already mentioned the treacherous ground of contractor's claims which come to a head at the settling of final accounts. The issuing of a Final Certificate with its explicit approval of all the works can too readily be seen as a prima facie case of negligence if anything subsequently goes wrong.

20.   Clearly some of these roles are readily adaptable to

the ways of Quality Assurances. Coordinating the design
team, applying the appropriate level of technology and
managing the flow of production information are all
functions likely to be improved by following clear and
explicit guidelines. There is nothing particularly creative
or contentious about these activities and one can imagine
them being increasingly systematized, automated and
computerised to the point where the production of large
buildings becomes increasingly uneventful. I think things
are going that way already though some would say much too
slowly. The effects for the designer will be two-edged in
terms of liabilities. On one hand procedures will be more
clearly and commonly understood and therefore easier to
follow. On the other hand any lapse will be equally clear
and therefore less easily forgiven.

21. Turning to his first and last duties it is harder to
see how QA procedures can help. Let us look at last things
first.

22. There are those who say that the designer should
cease to be involved in the building enterprise when he has
provided enough information to enable others to build his
design. The recent growth of management contracting is
based on that thesis. The result for the designer as
businessman would be beneficial, but for the designer as
artist the result could be disastrous.

23. In the Soviet Union in 1986, I met architects who had
had nothing to do with the building of their designs, that
being the domain solely of another part of the bureaucracy.
I remember visiting buildings of excellent concept and
careful design detail which had been literally destroyed by
a travesty of uncomprehending and insensitive construction.
The architects showing me around their buildings were almost
tearful in reminding me that they had no responsibility for
- in fact were not even allowed on site during -
construction. The buildings suffered immeasurably. Clearly
design sense and care must follow through to completion.
Stopping designing at tender stage is like lifting your head
when the golf club hits the ball and the results can be just
as wayward.

24. So assuming the designer is to continue to be
involved to the end of the contract, can QA help? Only, I
suspect, if the whole industry and especially contractors
submit to the same criteria of assessment. Good building is
a partnership of good design and good procurement methods.
As in history, designer and builder must again find and use
a common language.

25. Finally, we have the front end of the activity, the
solitary, initiating, conceptual act of synthesis. How can

QA assessment apply to or assist such a deeply personal act? Honest designers will admit that the process is not as mystical as might at first appear. Building types can be studied systematically, there are universally valid rules of scale and form and a wealth of precedent to provide a network within which ideas can be evaluated. It has happened for generations in the critiques in schools of architecture. It is the chief learning method of the art of architecture and it is open to systematic improvement, but only within a framework of great flexibility and sensitivity. Any crudeness of approach here will quickly stifle creative innovation.

26.   In today's architecture innovation is at a premium. Design is now a very important part of the commerce of ideas that characterises our age. Building design is more varied than ever and therefore more competitive. To compete in design means to innovate. To innovate is to take risks. To make the risks explicit to a client is responsible. To fail to do so is reprehensible.

27.   If QA can make designers and their clients more aware of the risks entailed in innovative design, and help them to confront the risks more bravely and responsibly it will be of great service to the building industry and might even inspire bolder and better design. The danger is that it may be applied too crudely by people who don't understand the true quality or nature of the activity to which they are trying to bring greater assurance. This is why Quality Assurance as a policy must start at the top of the industry where both its problems and its great potential can be seen most clearly.

Discussion on Papers 5 and 6

MR GRIFFITHS (Paper 5)
In 1986 consulting engineers notified more than 100
potential claims for every 100 policies. In the same year
41 design consultancies out of every 100 world-wide were
claimed against: the 1960 figure was 12. For every £100 of
premium they took from architects in 1986 insurers now
estimate they will have to pay out £250. The average 1986
claim cost insurers US$190 000. It cost the defendant firm
an 'excess' averaging US$25 000 and non-revenue-earning
time. Of the total payout, 70% went to the cost of defence.
  Faults in design concepts and parameters are the biggest
individual cause - 27% of all claims in 1985-86, rising to
33% in 1986-87. The main problem areas were steel frames,
foundations, architectural details including movement joints
and paving, and structural elements such as stairwells,
floors, roofs and walls. The sources of claims were
predominantly commercial buildings of all sorts, housing
developments and industrial buildings. Contractors'
insurances too have a long record of heavy claims.
  With figures like those it is not only the insurers who
will welcome measures to improve the risk, especially if
those measures reduce the number and cost of claims. If
adopting BS5750/ISO9000 cuts the number and cost of claims,
then insurers will be able to reflect it in the premiums
they charge.
  My paper aims to put QA into its wider framework from the
viewpoint of an insurance broker who specializes in
professional indemnity for construction professionals. This
contribution concentrates solely on QA, as it affects the
consultancy professions.
  The first step along the quality road is one which would
have been taken long ago - quality management. In smaller
firms it is probably quite informal, the boss makes a daily
round of the design office, checking as he goes. He
probably secured nearly all the work himself and therefore
knows all the firm's jobs personally. The kind of quality
management becomes impractical as the firm gets bigger. So,
in an effort to set and achieve uniform standards, the
partners will produce a procedures manual and check-lists.

The firm will then have started along the QA route.

To adopt BS5750, stage 1 the procedures manual is refined and extended. A quality manual is prepared and it is then called the quality system. The firm then starts to produce quality plans for its projects, and systems are updated so that records can be retrieved for proof that the working methods laid-down have actually been applied and that the checks specified were run.

Insurers will probably welcome it if the first stage of BS5750 is adopted: it indicates that a firm approaches projects consistently and applies checks to nip problems in the bud. Clients can then be more confident that whenever they appoint the firm it will perform consistently, even if it can not formalize inspiration!

Care is needed not to go overboard in the claims a firm makes; a consultant's legal duty is to bring to bear the reasonable skill, care and diligence of a competent practitioner, and that is the basis on which a consultant is insured. If first-stage QA has been adopted it should be considered as a management aid: any benefit to a client is a spin-off. A consultant could tell him simply that it helps one to manage the practice in a more structured and consistent way. Taking QA to stage 2 involves an external quality auditor, and if a consultant is working in sensitive areas like the nuclear industry and others which directly bear upon the health and safety of millions, I don't doubt that insurers would be comforted by knowing that there was an external quality auditor to satisfy.

The fact that a testing and inspection policy is agreed with the client at the outset and an undertaking is made to give him the test and inspection records might be helpful. The client is doubtless represented by someone himself highly qualified and expert in the particular field; if the consultant and the representative have agreed a testing and inspection policy at the outset it will not be easy for the client to come back later and accuse the consultant of negligent oversights - at least in the testing and inspection area. For a start a consultant should consider the liability of the Third Party auditor for his interpretation of BS5750 and its application in the practice - what does the agreement say on the topic? What rights might an injured Third Party have against that auditor? I have no ready answers but I do commend the thoughts to consultants for their attention.

Is a practice undertaking to reveal parts of its own inner workings which might be turned around and used against it? If it can be shown that a firm has failed fully to comply with its own QA manual and procedures does that, of itself, prove it has been negligent and in breach of contract and give a right to damages? If so how would the damages be quantified (especially if there had been no loss or damage)? There is a possibility that stage 2 QA could make a firm's legal position worse than it was before it went for QA. It

is of interest that one of the biggest companies represented
at this conference and one with a first class reputation for
quality, has apparently adopted stage 1 QA and is ready
formally to adopt stage 2 at short notice, but has decided
not to do so voluntarily at this juncture.

Newspapers reported in December 1988 that a consulting
engineer in Virginia is contesting fines of $17 000 for
allegedly violating safety and QA procedures after the
collapse of a 2000 ft television mast for KTVO. Among other
things the consultant is charged that he failed to maintain
an adequate QA programme to ensure the original structural
components met the specification. I expect he would not
have got the job at all if he did not operate QA, but it
looks like QA could land a consultant with not just a
liability in civil law for a breach of contract, but with a
criminal record as well, at least in the USA.

What is the position regarding the quality manager's
discrepancy reports? Are his audit reports of deviations
from the firm's quality plan open for 'Discovery of
Documents' in a law case, or can steps be taken to protect
them? Again, even if discrepancies represented a breach of
contract how would they be quantified for an award of
damages?

By offering itself as QA-accredited is a firm inferring
that it checks and reviews everything that leaves its
office? Or does it take positive steps to make it clear
that checking is done by sampling?

Is QA going to cause delays? Delay claims against
Engineers are already significant, but perhaps they are
preferable to even worse claims had one pressed on
regardless. Will there be significant cases of re-design
following QA checks, and, if so, will they be taken as
evidence that a firm was negligent, or accepted by the
client as a necessary stitch in time?

Does a consultant make sure the practice is engaged on a
Skill, Care and Diligence basis - which is what insurers are
entitled to expect? Is care taken to avoid giving a general
contractual commitment to carry out the QA procedures? They
are after all internal management systems, and ought not to
be elevated to contractual obligations to clients unless
there is a compelling reason and the client insists - for
example when work is undertaken with the nuclear industry.

As to calculating insurance premiums for QA-accredited
firms it is the case that some insurers, particularly in the
USA, claim to give discounts for QA and indeed for peer
review - 5% discount is typical. However, it is not clear
what the 5% is taken from.

Theories and hope may guide some insurers in their first
stab at fixing their premium rates, but the ultimate guide
is there on the bottom line: the claims experience. Those
looking to insurers for their pension will surely wish to
see them prosper; professional indemnity historically runs
at a loss ratio significantly in excess of 100% and it is

really only a combination of investment skill and optimism
that keeps insurers in the professional indemnity marketplace.

It is early days yet to see whether QA is affecting
insurers' financial results since professional indemnity
claims usually take seven to nine years after practical
completion to emerge and be settled - that may sound like a
very long time but patience pays off. In due course QA may
show good effects, or perhaps downsides, which have not yet
been foreseen.

MR D. A. HEATH, Department of Transport, Bristol
In what way should the client go along with the QA
consultant: in spirit, by involvement in the firm's QA
scheme or by having QA itself?

MR V. J. W. HOAD, Sir William Halcrow & Partners Ltd
Mr Heath enquired what part clients had to play in quality
assurance. It is vital for clients to ensure that they
issue and receive communications from designers in a very
disciplined and timely manner, otherwise the whole system
could be broken. It is vital for both sides to be fully
aware of all data that is exchanged.

7.   What does 1992 hold for the materials producer?

K. NEWMAN, BSc, MICE, FICT, British Cement Association

SYNOPSIS. With less than four years before the completion of a
Single European Market, the UK Materials Industry by and large
understands what is involved, is aware of the potential threats
and ready to take up the opportunities which are being created.
This paper, a personal view by someone who has been involved in
European Standardization for over 10 years, discusses progress
in the harmonization of Standards, Certification and Testing
procedures, and speculates on their eventual impact once the
Single Market is in being.   Although there are wide ranging
implications, to a large extent UK construction products
companies are already well prepared.   Others' may find that, in
the event, their activities may not be too greatly affected.

INTRODUCTION
As always, now and in 1992, the materials producer must supply
materials, products and components which are 'in accordance
with the standard or specification', i.e. of the required
adequate quality, at a reasonable price and so provide value
for money. As far as the Client is concerned, the product must
be 'fit for the intended purpose and perform satisfactorily in
use'.   However, this presumes that products of the necessary
quality have been specified in the first place.

The obvious opportunity presented by the opening up of the
Single European Market in 1992 is that it will provide access
to some 320 million people in the European Community (EC), a
market which is larger than that of the US or Japan.   The
removal of barriers to trade will enable easier marketing of
products through harmonization of standards and practices, more
opportunities for contractors to tender for public sector
projects and allow professionals to practice more easily
throughout the EEC.

The main problems are concerned with the harmonization of
practices with which UK producers are either not familiar or
prove to be to their technical or commercial disadvantage.
There will inevitably be the threat to UK suppliers,
contractors and professionals from increased competition in
their home market from their counterparts in the rest of
Europe. The considerable investment already seen in South East

England and the present construction boom is a clear indication of this.

What are the rules and regulations being developed to enable the Common Market to operate and how will they affect the materials producer? Let us first review briefly the EC legislation and specifications which are being drawn up.

THE BACKGROUND

For the past 15 years or more, a framework of instruments for the removal of technical barriers to trade and the harmoniz- ation of standards and regulations has been rather painfully and, let it be said, unsystematically erected. With the approach of 1992, there has been an acceleration of activities but many things are still far from clear. However, there is now the basis of a system which will govern the trade in construction materials and products in Europe, but which will inevitably be subject to much refinement and consolidation before the end of the century.

The main aim of the Single Market has been the removal of technical barriers to trade, i.e. those differences in product regulations, standards or compliance procedures which can hinder or even prevent a product acceptable by one Member State from being traded freely in another. The two-pronged attack on technical barriers has comprised:

- Harmonization of technical regulations and standards according to the new approach adopted in 1985 by the European Council of Ministers
- Mutual recognition between Member States of each others' testing and certification procedures and results

At the beginning, technical harmonization was a slow and cumbersome process with the Commission itself trying to draw up precise technical criteria by which the requirements of the directives were to be satisfied. This led to years of discus- sion and argument over minute technical details for individual products.

This difficulty of drafting European or International standards for the multitude of construction products has, to some extent, been overcome by the new approach introduced in the Commission's White Paper of June 1985 in which Directives are to be prepared covering wide product categories. This 'new approach' to technical harmonization and standards is based on EC standards for health and safety which afford all Europeans with an equally high level of protection and yet leave manufacturers, whose products meet the standards, freedom to use their own manufacturing and design traditions and skills.

In these Directives, harmonization legislation is to be restricted to laying down essential safety and other relevant

72

requirements to which the products must conform. The Commission has given a mandate to the European Standards organisations CEN (European Committee for Standardization) and CENELEC (European Committee for Electrotechnical Standardization) to draw up the technical specifications needed for producing and marketing products conforming to these essential requirements.

For this approach to work, there has to be a clear different- iation between those materials, products and components where harmonisation is necessary or advisable and those which can be covered by mutual recognition of national standards and regul- ations. This distinction is, and will be, a matter of dispute and we shall see much manoeuvring and bargaining before, if ever, we shall have true and complete 'Europeanization' of construction products. It must be remembered that once a harmonized standard has been agreed by a majority vote, any conflicting National Standards will have to be withdrawn.

CONSTRUCTION PRODUCTS DIRECTIVE
In the White Paper, the Construction Industry was earmarked as one of the priorities for the new approach to technical harmonization and standards. Other Directives under discussion which affect the Construction Industry cover public works and supplies, mutual recognition of professional qualifications, product liability and minimum health and safety requirements for work places. However, for the materials producer, the Construction Products Directive will have the major impact.

A draft of the Construction Products Directive was submitted to the Council early in 1987 and the text is being negotiated through the Internal Market Council. Its aim is to provide for the free movement, sale, and use of construction products which are fit for their intended use and have such characteristics that structures in which they are incorporated meet certain essential requirements. These essential requirements are similar in style but wider in scope to the functional requirements of the Building Regulations in force in England and Wales. They relate to mechanical resistance (stability), safety in case of fire, hygiene and health, environment, safety in use, protection against noise and energy economy.

Products will be presumed to be fit for their intended use if they bear a CE conformity mark showing that they comply with a European standard or a European technical approval. Where European documents of this sort do not exist, relevant national standards or Agréments may be recognized at Community level as meeting the essential requirements. If a manufacturer chooses not to make a product in conformity with the specification, he has to prove that his product conforms to the essential requirements. One method of verifying conformity is by third party certification.

NECESSARY STANDARDS AND TECHNICAL APPROVALS

The key factor in this new approach, is the preparation of European standards by an approved European standards body on the basis of a mandate given by the Commission following consultation with a Standing Committee of representatives of the Member States. The European Standards body will usually be the CEN or its sister organization, CENELEC.

European technical approvals will be issued by approvals bodies designated for this purpose by the Member States in accordance with guidelines prepared by the recognized European body comprising the approvals bodies from all the Member States. This European organization will be similar to the existing European Union of Agrément (UEAtc), on which the British Board of Agrément is the UK member.

The mandates issued by the Commission to the European standards and technical approvals bodies are to be based on a series of technical documents which are subject to approval by the Standing Committee. These will define the necessary essential requirements for the products concerned and bridge the gap between these and national regulations and the European specification. For some years the Commission has been preparing a series of structural Eurocodes which relate to mechanical resistance (stability). Preparation of documents relating to the other essential requirements will also be carried out at the request of the Commission and Standing Committee.

TESTING AND CERTIFICATION

The success of the new approach is dependent on the ready acceptance in one Member State of certificates and test results produced in another Member State as proof of conformity to the essential requirements laid down in a Directive. This will also involve the mutual recognition of certificates and test results in areas where harmonization is not appropriate. In the past each Member State has developed its own approach to these subjects.

Since January 1986 there has been considerable activity in the development of an EC policy on the harmonization of testing and certification practices. This is seen as the means of creating the right environment for mutual recognition, thereby removing the need for goods and services to be re-tested and certified as they cross national frontiers. General criteria for the running of and assessing the competence of certification bodies, testing laboratories and accreditation organizations have already been drawn up and CEN/CENELEC have been mandated by the Commission to produce European standards based on these.

As a next step, the Commission embarked on extensive consultation to develop an EC philosophy towards certification and testing. As part of this exercise they have examined procedures in use in the Member States, sector by sector, in order to establish which can be regarded as producing equivalent results from the points of view of health and safety. As a follow up to its consultation paper, the Commission has produced two further documents on testing and certification in which they have put forward preliminary ideas for a suitable framework within Europe. Although these ideas still need considerable refinement, the documents have nevertheless proved a useful basis for further discussions with the Commission.

At a symposium in Brussels in June 1988 to discuss testing and certification policy matters, there was consensus on the need to overcome the technical barriers to trade caused by goods having to be re-tested and re-certified when they crossed national frontiers. There was broad agreement that the way forward was through mutual recognition of test results and certificates. There is also a general feeling that this can best be achieved by means of market-led recognition agreements covering identified sectors coupled with an organisation of some sort to provide technical and administrative support.

PRESENT SITUATION

So we will have a Construction Products Directive, CEN Standards and European Technical Approvals, and mutual recognition of certification and procedures - these are the instruments which are being developed. Are they complete? Can they be used? Will they make much difference to the present situation regarding the import and export of construction products? At the risk of being cynical, at the moment the answer to these questions is NO.

The present construction boom in South East England has put great pressure on UK product suppliers. Materials, products and components are being imported from all over Europe - with not an EN Standard or Certification Scheme to be seen anywhere!

It must also be remembered that, for the most part, construction is a local affair concerned with meeting local needs according to local rules. Whilst the Single European Market may be aimed at removing technical barriers to trade, it cannot remove the age-old 'Environmental barriers'. Single Market or no Single Market, the weather, temperature and humidity conditions in Helsinki will never be the same as in Athens, and construction products which perform perfectly satisfactorily in the Mediterranean may be completely unsuitable in Scandinavia.

Codes and Standards. The Client requires the assurance that his structure, building, etc. will meet <u>his</u> essential requirements and be suitable for the purpose. He relies on his design team of architect and engineers to ensure the integrity of his structure and the fitness for purpose of the materials which are used in it. Harmonization of the structural integrity of buildings is being achieved through the Eurocodes, or Codes for Construction covering:

1.   General principles
2.   Concrete
3.   Steel
4.   Composite structures
5.   Timber
6.   Masonry
7.   Foundations
8.   Seismic
9.   Actions/loadings

These Codes are basic technical reference documents which can be used to demonstrate compliance with National Building Regulations or to justify an alternative design in connection with a tender invited by a Public Works Authority under the Public Works Directive. The Eurocodes cannot be implemented until there exists a large family of supporting product standards, the preparation of which is the responsibility of CEN. At the moment there are more than 50 Technical Committees tackling a wide range of construction products. A European Standard (EN) is drawn up on a consensus basis and adopted through a weighted majority voting procedure. An adopted standard must be implemented in full as a National Standard in each CEN country, and any standard which conflicts with it must then be withdrawn.

The considerable difficulties in achieving a consensus, has led to the attempts to produce European Pre-Standards (ENV) similar to a BSI Draft for Development. The fact that after more than 10 years Standards for cement and concrete have still to be produced, even as an ENV, indicates the difficulties involved. So the questions must be asked, is European harmonization warranted or even necessary for construction products? In many cases probably, in some possibly, but in others not at all.

Compliance with the Construction Products Directive. The basic problem is that, unlike many consumer products for which the Directive approach was designed, construction products are not 'end user products' but are intended for incorporation in structures. Therefore the 'essential safety requirements' in the Directive are related to the end product which is the construction works. The Directive requires that all construction products must have characteristics which will not prevent the building from meeting the essential requirements in the Directive. A manufacturer will be able to demonstrate these characteristics by one of several different routes:

i.   Compliance with the harmonized European Standard (EN) or
ii.   Compliance with an approved National Standard where no EN exists, or
iii. The holding of a European Technical Approval (e.g. Agrément Certificate)

It is possible that even attestation of compliance with the Directive by the manufacturer, may be permitted. Certainly manufacturers who can demonstrate the fitness for purpose characteristics for their products required by the Directive will be entitled to use the CE mark.

Agreed Test Methods - The Key. Meanwhile, as the debates continue, the designer still wishes to know which products are available, detailed information on their quality and how they perform, and assurance that those used will be in accordance with the specification. He will only be able to achieve this if he can define clearly the quality grade of the products he requires in performance terms. For this reason, the main thrust of European activities should be directed at agreeing test methods. This has certainly been the case as far as the proposed EN standards for Cement prepared by the CEN/TC 51 Committee is concerned. Agreement has already been reached on methods of testing published as EN 196 which will provide a firm basis for the early approval of an ENV for the specification of cement, ENV 197.

Fitness for Purpose. A major potential loophole in the Construction Products Directive and European standardization is in identifying the suitability or fitness for purpose of products for different uses. In the proposed European Standards for Concrete ENV 206 the required qualities of concrete are defined in terms of exposure classes or environmental conditions. These are expressed in very general terms and could lead to the demand for concrete mixes of qualities which may or may not be suitable throughout Europe. Concrete structures in Scandinavia are certainly subjected to very different conditions to those in the Mediterranean.

Free and Fair Competition. The material supplier requires a clear specification of the qualities required of his products in performance terms and freedom to satisfy the specification. Furthermore he requires an assurance that there will be fair competition in the market place. Again one of the problems is that Member States will not be able to erect regulatory or technical barriers relating to the essential requirements against products which have the CE mark. Although there is a safeguard clause which enables a Member State to prohibit the sale of incorrectly marked products, there is a very real risk that the CE mark will become synonymous with the lowest quality.

Against this background there are the further relatively unknown fields of mutual recognition of Certification and Testing procedures which is certainly likely to be the next major battleground.

At this time, it appears that the safest road to sensible harmonization of Construction Products is:

i. To approve testing methods which allow the determination of different qualities in performance terms

ii. To present the qualities of products as a range of grades or classifications so that the Architect and Engineer can choose the right product for his purpose and through appropriate testing and certification be assured that those used will be in accordance with the specification.

CONCLUDING REMARKS

For the materials, products or components supplier who has yet to be involved in the European scene, this somewhat cursory survey may cause him considerable alarm, and the feeling that he would much prefer not to be involved at all! However, the benefits of the Single Market are worth remembering. It will create a new domestic EC Euro Construction Market of some £250 billion (compared with the UK £35 billion), which is 40% of the World market. The value of building materials supplying this market alone is nearly £75 billion. The economies of scale will affect not only production but also more importantly, research and development. All companies should re-assess their marketing policies to meet increasing competition. New products can and must be evolved as their production becomes more economic.

For building products, the main barriers to trade at the moment are the complicated National regulations and individualistic certification procedures. France and Germany are the most demanding with their requirements for importers, whilst in the UK any restrictions are rarely mandatory. The benefits to companies wishing to sell in Europe is that they will no longer have to manufacture goods to a wide range of specifications nor having to wait years for certification by the importing country.

Material producers should be aware that their competitors, particularly in France and Germany, are working flat out to establish rules and regulations which are to their benefit. All those who have been involved in European harmonization will confirm that the easiest path into Europe is to be actively involved in drafting and commenting on the standards and regulations that are currently being prepared. In the European harmonization jungle, it is essential to be 'quick on the first draft', but to do this one must be actively involved. All materials producers should consider their current activities against the checklist given in the Questionnaire. If your company can answer Yes to most of the questions, then you are certainly well ahead in preparing yourself not only to combat the threats but more importantly to make the most of the opportunities of the Single Market.

SOME USEFUL REFERENCES AND CONTACTS

1. "The UK Construction Industry and the European Community".
    National Council of Building Materials Producers and
    Building Magazine.

    A comprehensive, concise handbook giving useful background
    to the subject. An essential European Directory and
    reference book.

2. Euronews Construction

    Periodic publication by Construction Industry Directorate
    (DoE). Regular information and Updates on EC Construction
    Industry matters.

3. Standards - General European Matters (Geoff Strawbridge,
    BSI 01-629-9000)

4. Construction Products Directive (Mr Oliver Palmer - DoE
    Construction Industry Directorate 01-276-6536 or Anthony
    Davies 01-276-6725)

5. Testing and Certification (Ian Pannell - DoE/CID
    01-276-6730)

6. European Work on Testing and Certification (Roger Brockway
    BSI 01-629-9000)

7. European Technical Approvals Work (Norman Garner, British
    Board of Agrément 0923 670844)

8. European Work on Testing (John Summerfield NAMAS
    01-943-6266)

CONSTRUCTION PRODUCTS SUPPLIERS CHECKLIST OF QUESTIONS

THE MARKET
1. What business is my company in and what is our UK market?

2. What is our market in Europe?

3. Do we already have associated companies, partners or
    business links in Europe, or are we exporters or potential
    exporters, or only concerned with competition from
    imports?

4. Have we reviewed our European policy for production and
    distribution, marketing and sales, R&D and investment,
    recruitment and training?

PRODUCT SPECIFICATIONS AND CODES

5.   Are our products covered by a British Standard or Agrément certificate?

6,   Is the British Standard a performance specification with different grades of quality?

7.   Is there a Eurocode covering our products and are we fully conversant with its implications?

8.   Are our products covered by a current CEN Technical Committee?

9.   Are we fully conversant with other European national standards covering our product range?

TESTING AND CERTIFICATION

10. Are the BS test methods for our products acceptable and satisfactory?

11. Are there agreed and acceptable European or ISO testing methods for our products?

12. Does our company operate a Quality Management System in accordance with ISO 9000 (BS5750)?

13. Are our products covered by a certification procedure in the UK (i.e. BSI, National Accreditation Council, Agrément)?

14. Are we certain that certification procedures in other European countries provide the same degree of assurance as those in the UK?

REPRESENTATION

15. Is my company a member of a UK trade federation?

16. Is our trade federation fully represented on BSI Committees?

17. Is our trade federation involved in BSI International Committees?

18. Is our UK organisation a member of a corresponding European trade organisation?

19. Is the Euro organisation actively involved/being consulted in the preparation of appropriate test methods and standards?

20. Are our staff involved in the drafting work of European Standards?

If your answer to questions 5 to 20 is Yes, then your company is fully prepared for Europe and 1992!

8. Harmonized design — a view of quailty assurance in the EEC

R. H. COURTIER, MA, FICE, W. S. Atkins & Partners

SYNOPSIS We are all well aware of the existence of the European Economic Community (EEC) and its concepts of harmonisation between the Member States. How real is this concept when applied to design? Is 1992 really going to be the big change forecast, or is it only a dream?

This paper examines through the eyes of a designer, what role codes and standards have in the design process and how the framework of documentation being set up by the EEC appears likely to affect it. Inevitably it is only today's view as the situation is developing at a fast pace.

CODES AND STANDARDS

1. Some universities still persist in considering engineering to be an 'art'. This is perhaps justified as civil engineering relies very heavily on empirical methods. For the same reason, the design process is susceptible to complex and disparate codification. The lack of purity in the science of civil engineering must be clearly recognised in any attempt at rationalisation of methods of design. We may be able to create mathematical models and convince ourselves that the purity of maths is real, but the slightest contact with construction practice soon demonstrates that the design process relies **entirely** on empirical correlations between the mathematical model and the prototype performance. It follows therefore that there are no single technical solutions in engineering.

2. Our social and economic culture demands that in the areas of performance, safety and environmental impact, some standards of acceptance must be established for any activity affecting other persons. Where any degree of complexity appears in the activity, the simplistic concept of quality control - i.e. does it or doesn't it meet stated criteria - is found wanting because of the range of possible circumstances to be considered. Quality assurance has been developed as the management system enabling appropriate levels of confidence to be established for any

product, however complex. In very simple terms, the essence of such a system may be set out as:

- Define characteristics of end product
- Define method of achieving characteristics
- Carry out method
- Audit compliance with method.

Each of the processes must be separately validated. The frequency of validation and stringency of audit will determine the degree of assurance conferred on the end product.

3. It is clear from the foregoing that assurance relies on a complex interaction of methods of working (codes of practice) with varying criteria (standards); our BSI Standards have over the years attempted the very difficult job of providing a framework to this process. Each Member State of the EEC has developed its own framework and very little has been done in the past to correlate these different frameworks. Indeed, there is little point as they are rarely, if ever, congruent even in philosophy.

4. The concept of 'harmonisation' of codes and standards must thus be accepted therefore as wholesale change - a new set of rules. There cannot be any illusion that it is some slight adjustment of clauses in existing codes. Fortunately the design process in the last 15 to 20 years has got used to a large number of changes in codes. This has created a measure of heart-searching amongst the older professionals and a problem for education and training. A further change is required to achieve harmonisation; more complete, more penetrating and, on the track record so far, requiring greater technical skills in understanding.

PRODUCTS IN DESIGN

5. For the obviously advantageous reasons of economy the design process embraces the concept of sub-products for use in the works. There is a very large range in the level of design and manufacture that may be involved in these sub-products varying from the lowest processing level of, say, gravel aggregates, to elements that effectively embody the same inputs as the finished works such as precast concrete frame elements.

6. The efficiency of the design process relies on the integration of the characteristics of sub-products with those of the final end product. If the construction process is also to be efficient there must be a limited interface between the design phase and the procurement phase; non-conformity at this interface will cause the need

of the sub-product's characteristics to be iterated back into the design process, with consequent delays. Such problems are commonplace with the integration of say, mechanical and electrical building services within civil works. It is well-known that not only does this process cause delay to the civil engineering but it also creates more difficult assurance procedures.

7. It can clearly be seen from the foregoing that the specification of products is intimately connected with the method of design; indeed many products not only themselves have to be based on compatible methods of design to those used for the works into which they are incorporated, but they condition that design also by their own detailed characteristics, such as the coefficient of thermal expansion, which may not be their primary characteristics. Most construction and building products are used compositely as an integral part of the works - in this respect they differ from many other manufactured items which have 'stand-alone' performance.

8. Although not always immediately apparent, the overall performance and safety of completed works is dependant on this necessary coherence of codes and standards. This coherence ranges from loads, partial safety factors, frequency of testing and test methods down to the instruments of test themselves. Our custom and practice in the industry is also highly dependant on familiarity with this coherence - it provides a guiding hand to efficiency.

STRUCTURE OF DOCUMENTS IN EEC

9. Having considered the philosophy of codes and standards the consequences of the Single European Act establishing a single common market by 31 December 1992 can be put in focus. Two groups of bodies influence this.

10. The first group involved are the four political bodies, whose structure is not dissimilar to that familiar in the UK. The principal authority is the **Council of Ministers** of the (currently 12) Member States, who ratify policy. They are advised and restrained by the **European Parliament** on matters of budget and in policy making. The equivalent of the Civil Service, creating policy documents and implementing them is the **European Commission.** The fourth body is the **Court of Justice** which provides judgement on compliance by the Member States.

11. For a good number of years the Commission has been attempting to produce a number of standard codes for construction - the Structural Eurocodes (the 9 proposed are

set out in the previous paper to this Conference by K Newman).

12. This process of drafting has proved extremely difficult and time-consuming and has now fallen foul of procedure. Until very recently, a Commission's Steering Committee oversaw the activities of a number of drafting panels, one for each of the Codes. The members of these latter panels were not representative of the Member States, although the Structural Eurocode Steering Committee itself had full representation.

13. It appears that this process and its parallel systems feeding via liaison engineers within the Member States and an editorial group, was not entirely acceptable. In consequence a new route is being established. This introduces the second group of bodies, who are voluntary and stand aside from the main political structure.

14. Most important of these organisations is the standards body CEN (Comitee Europeen de Normalisation) and its counterpart for electrical technology standards, CENELEC. CEN is comprised not only of the national standards bodies of the 12 EEC Member States but also the 6 additional EFTA states.

15. CEN has now been mandated by the Commission with the role of producing the Eurocodes; exactly how this will be done however, has not yet been made completely clear. It it is expected that the codes will go through the national standards bodies for comment even though input has been obtained from this source already. Progress in this area is now seen as delayed, although not necessarily in the sense of having lost ground. Nevertheless it may see a shift from the predominantly academic input to the codes so far seen.

16. The main activities of CEN concern the management of the many Technical Committees (TCs) of which there are several hundred at present. These TCs have the same characteristic names of many existing BSI committees, providing both narrow views, such as 'TC105-Valves and Fittings to equip Radiators', and more general aspects such as 'TC127-Fire Safety'. The output of these committees is to create agreed standards documents, falling into one of three basic types, as follows:

EN - European Standards
HD - Harmonisation Documents
ENV - Pre-standard Documents

The latter ENVs are similar to BSI's PD documents, issued

for development, whilst the Harmonisation documents cover standards that cannot be fully implemented because of impracticabilities associated with specific national differences (such as climate).

17. At first sight this appears all to be entirely in accordance with the needs of the philosophical requirements, viz: Structural Eurocodes setting out the basic method of design, and European Standards the particular acceptance criteria. This might be so when the majority of documents have been agreed, but raises a question during the transition stage.

18. It must be understood that the codes and standard just referred to are voluntary documents; CEN only has the mandate to create them, not to impose them. Members are however bound to adopt ENs or HDs once agreed by a new weighted voting system, and as a result withdraw national standards in conflict. Two major problems thus appear to thwart the concept of the Single European Act. Firstly the question as to what part of any standard is to be mandatory and the second how is mutual acceptance to be obtained as to whether those mandatory requirements have been met. These points have been considered by the Commission.

THE CONSTRUCTION PRODUCTS DIRECTIVE 9890/88

19. Construction Products, which the Commission defines as meaning products which are produced for incorporation in a permanent manner in construction works, including building and civil engineering works, are subject to a directive that has recently been ratified by the Council of Ministers. One of its clauses sets the scene very clearly.

20. Article 2 Clause 1 directs that Member States shall take all necessary measures to ensure that products which are intended for use in construction works, may be placed on the market only if they are fit for this intended use, that is to say they have such characteristics that the works in which they are to be incorporated, assembled, applied or installed, can, **if properly designed and built**, (author's emphasis) satisfy the essential requirements when and where such works are subject to regulations containing such requirements. The 'essential requirements' are set out as covering

- Mechanical Stability
- Safety in case of fire
- Hygiene, Health and the Environment
- Safety in use
- Protection against noise
- Energy economy and heat retention.

Products satisfying these requirements may have a CE mark designation, by which they can then freely enter the market.

21. It is immediately evident that these requirements have a quite different meaning when applied to the products themselves as opposed to the completed works. The caveat 'if properly designed and built' controls the interrelation. It is at this point that the problems facing the designer begin to multiply. If compliance with a CEN standard or other transitionary document was a presumption of meeting the essential requirements, and this former compliance was both a necessary and sufficient condition of the presumption, all would be well. However, the important let-out must be included for the transition or innovative situation where no standard exists. Thus this route of compliance becomes only one of several sufficient conditions. Another route created to satisfy compliance is that of 'technical approval', which may apply to a product which has not harmonised or national standard. It is defined as 'a favourable technical assessment of the fitness for use of a product for an intended use, based on fulfilment of the essential requirements for building works for which the product is used'. This immediately raises the question of mutual recognition of conformity to the essential requirement.

THE COMMISSION'S POLICY ON TECHNICAL SPECIFICATIONS TESTING AND CERTIFICATION - CERTIF 88/10

22. Considerable energy has been devoted since 1987 to the solution of the problem of mutual recognition of testing and certification, it being realised that to ensure free circulation of goods and more so, to create an internal market based on a single homogenous technical environment, such recognition is essential. To that end a European Organisation for Testing and Certification is being set up to fill the void previously existing.

23. The most recent related document, Certif 88/10 (29 November 88), provides the current policy of the Commission in this area.

24. Whilst setting out the details of the seven methods (refer to Table A) of assessment of compliance with the essential requirements and EC mark conformity, this document embraces both the route of assessment of essential requirements where no standards exist, and the non-uniqueness of technical solutions to conformity. In these latter cases attestation of conformity is left to the producer, albeit only after depositing design documentation with an approved body.

25. The conclusions to be drawn from this policy

Table A

CONFORMITY ASSESSMENT PROCEDURES IN COMMUNITY LEGISLATION

A) EC declaration of conformity	B) EC type examination	C) EC declaration of production conformity (type 1)	D) EC declaration of production conformity (type 2)	E) EC declaration of production conformity (type 3)	F) EC verification (type 1)	G) EC verification (type 2)	H) EC declaration of design and production conformity
- Design documentation at the disposal of notified body Manufacturer - ensures conformity with design documentation - declares conformity and affixes the CE mark	Manufacturer submits to notified body - design documentation - type, if necessary Notified body - examines design documentation - tests type, if applicable - Issues type approval certificate	Manufacturer - ensures conformity with approved type - declares conformity and affixes the CE mark	Manufacturer - Implements Quality Assurance (QA) for production and final testing - declares conformity - affixes the CE mark - Is subject to surveillance of QA by notified body	Manufacturer - takes Internal measures for product control - declares conformity - affixes the CE mark - Is subject to surveillance of products by notified body	Notified body - verifies conformity of each product with the type as described In the type approval certificate - Issues certificate of conformity - affixes the CE mark	Manufacturer submits product(s) Notified body - verifies conformity of each product with the requirements of the directive - Issues certificate of conformity - affixes the CE mark	Manufacturer submits design documentation Manufacturer - Implements QA for design control - Is subject to EC surveillance of QA by notified body Manufacturer - Implements Quality Assurance for production and final testing - declares conformity - affixes CE mark - Is subject to surveillance of QA by notified body

87

document is that whilst the route to mutual recognition of testing is mapped out, the passport to interborder trading - the EC mark - will have very little relevance to the designer without a great burden of research. During the agreed transitionary stage whilst standards are still being issued, the advantage of having standards to limit the interface between detailed design and procurement as set out earlier is potentially lost. Instead the existence of non-standard items, all deemed to have equal status in the market place, because they satisfy the lowest common denominator - the essential requirement - will have to be dealt with.

CONCLUDING REMARKS

26. Looking carefully at the management of the design process it is apparent that efficiency is to a large extent created by limiting the number of decision paths to be considered. Successful assurance of quality also benefits from this limitation.

27. Innovation involves many more decisions than the mere incorporation of a prefabricated product. Economics thus demands that we limit our innovation to where it is important and, eschew the proverbial reinvention of the wheel. The existence of standards is a powerful tool in achieving this economy. Standards also significantly improve efficiency in the verification stage because objective criteria remove arbitrary disapprovals.

28. The construction market in the UK is not as well organised in the use of objective criteria as it should be. Too many vague specifications are used and even British Standards fail sometimes. For example the 'Specification for aggregates from natural sources for concrete BS882 : 1983' states for the contaminant material clay - i.e. dirty aggregate.

> 'As no test is available at present for determining clay as an adherent coating or as lumps in aggregate, no limits have been specified at present in this standard'.

29. Such lack of specification means that other methods of assurance must be relied on, whether by the supplier or engineer. It is clear that assurance of quality in the UK market relies heavily on 'previous satisfactory use'. This is not to be criticised as, properly assessed, it may be an extremely good test. It does, in the face of vague specification however, point to the change that 1992 will inevitably bring.

30. On the presumption that the seven methods of

assessment set out in Table A will govern products on the market; experience in use will not be a satisfactory, or, in those areas of regulation, acceptable basis of selection. Designers and procurers are going to have to get used to properly specifying their needs and to require the appropriate testing - they just won't have a sufficient experience base to rely on previous use.

31. Until a very large proportion of EN standards are in place only one route of protection of the designer's duty 'to properly design' appears to exist. The discipline imposed by a quality system replaces the vagaries of approval by the engineer with objective acceptance criteria whilst retaining contractual rights to reject or waive non compliance. In this way necessary documentation can be specified in the procurement phase. All EC marked items will have attestation documentation even if 'self-assessed', so this seems a viable approach.

32. Both the designer and the procurer will certainly have to increase their awareness of the market, but that is inevitable. If the political dream is a market of 320 million freely trading, whilst at the same time providing an element of a protectionist economic counter-attack to Japan and USA then there **will** be many more items on the market. Not all of them will be neatly packaged to European Standards. A change in the approach to procurement will follow.

33. This paper has tried to place in context the political ideal of the free market and the pragmatic needs of the designer who has to get the job detailed and specified efficiently. 1992 is still a long way off and there is plenty of time to get into shape for that future scene, but it will undoubtedly require effort to do so. It would not be unduly cynical to record progress against the Government's time-table for metrication in the early '70s - more than fifteen years later we have not achieved this.

34. The Eurocodes are likely to be adopted nationally as ENs and will require considerable digestion - at present they have plenty of wickedly academic passages. But above all there will be the need to properly understand, and by that understanding, properly specify, materials and products. Without the discipline of quality assurance procedures this will not be possible. The political pressure to create the market is strong. The key question is whether we, the industry, can get it right.

Discussion on Papers 7 and 8

MR G. H. COATES, <u>Sir Alexander Gibb & Partners</u>
Clearly both Mr Newman and Mr Courtier have given a great
deal of their time to working on Eurocodes and to Euro
Committees and, indeed, Mr Newman said he has been engaged
for 10 years doing just this. Yet the necessary work in
connection with Eurocodes and standards is only beginning.
Several individuals and the firms or organizations that
support them are already beginning to feel that the cost to
them in time and money is becoming more than they can stand.
Should pressure be exerted for Government to provide funding
for any of this work in order that the UK may stay ahead or
abreast of the other countries?

MR B. O. GRIFFITHS
Do the authors of paper 7 and 8 foresee European Conditions
of Engagement and/or European Conditions of Contract
supplanting those the UK has used for many years and of
which the precise meaning at law has been learned at great
cost?

MR COATES
In answer to Mr Griffiths, both British Consultants Bureau
and the Association of Consulting Engineers are currently
looking at, and commenting on, the draft Directives for
Procurement. It is also being looked at by CEDIC, which is
the European organization of national associations of
consulting engineers and as far as possible these bodies are
co-ordinating. It is important that due attention is given
to such documents and the lesson was learnt quite late in
connection with the Procurement of Works document used by
the EEC in connection with the disbursement of EDF funds —
this is the Fund set up by the EEC and available to
signatories of the Lome Agreement who are all developing
countries. The countries who had originally been used to
British Conditions of Contract suddenly found that they were
faced with re-learning all the knowledge that had been
transferred to them and came together to attempt to persuade
the EEC to change the document, but they were extremely late
and their effectiveness was thereby diminished.

MR J. R. T. DOUGLAS, Federation of Civil Engineering
Contractors
The FCEC, together with the Building Employers'
Confederation, is studying possible European forms of
contract, and also the Public Works directive and those
covering compliance and excluded sectors. They are also
considering European company law and the arrangements
proposed for mergers and competition. This work is being
carried out in the UK and in Europe with FIEC.

MR K. A. L. JOHNSON
Conditions of employment in Europe for both professionals
and operatives differ widely, not only with respect to
safety but also hours of work, social facilities and so on.
What is being done to rationalize these conditions to ensure
fair competition?

MR DOUGLAS
Also being studied inside the FIEC is the whole question of
employment legislation, especially in the light of the
encouragement to European trades unions given to the British
TUC at its 1988 Bournemouth Conference. FIEC sees the need
to influence European construction employers, some of whom
are accustomed to a more socialized regime, to take into
account the views of those from a more free enterprise-
oriented background.

MR COATES
There is certainly a lot of activity spread very widely in
connection with 1992 and in particular the Directives for
Procurement of Services and Works. Indeed, I have heard
mention that there are as many as 26 organizations concerned
with the construction industry, who have panels or Euro
Committees. There is a great need for sufficient coordination in order to make sure that wasteful duplication
of effort is avoided.

MR NEWMAN
Mr Coates is quite right when he says that industry, through
trade associations or individual firms, has already provided
considerable support to Euro-Committee work and this will
certainly increase. In fact, Government departments are
becoming aware of the need for providing funds for
individuals and experts who do not have an organization or
company to support them directly. However, there is also an
allied problem of finding sufficient independent
representatives, e.g. from DoE or BRE, to suupport the
growing number of committees which are required to achieve
harmonization in the required time scale. The cutbacks in
recent years in Government departments and research
establishments has led to increasing pressure on staff time.
Government must be made aware of the weakening effect this
can have on UK representation at the harmonization table.